Putting Essential Understanding into Practice Series

Putting **Essential Understanding** of

Number and Numeration into Practice

in

Prekindergarten–Grade 2

Karen S. Karp
Johns Hopkins University
Baltimore, Maryland

Barbara J. Dougherty
University of Hawai'i
Honolulu, Hawai'i

Karen S. Karp
Volume Editor
Johns Hopkins University
Baltimore, Maryland

Barbara J. Dougherty
Series Editor
University of Hawai'i
Honolulu, Hawai'i

NATIONAL COUNCIL OF
TEACHERS OF MATHEMATICS

1906 Association Drive, Reston, VA 20191-1502
(703) 620-9840; (800) 235-7566; www.nctm.org

Library of Congress Cataloging-in-Publication Data

Names: Dougherty, Barbara J., author. | Karp, Karen S., author.
Title: Putting essential understanding of number and numeration into practice in pre-K–grade 2 / Barbara Dougherty, series editor, author ; Karen Karp, author.
Description: Reston, VA : National Council of Teachers of Mathematics, [2019] | Series: Putting essential understanding into practice series.
Identifiers: LCCN 2018055991 (print) | LCCN 2018060620 (ebook) | ISBN 9780873539166 (ebook) | ISBN 9780873537162
Subjects: LCSH: Number concept–Study and teaching (Preschool) | Number concept–Study and teaching (Primary) | Numeracy--Study and teaching (Preschool) | Numeracy–Study and teaching (Primary)
Classification: LCC QA141.15 (ebook) | LCC QA141.15 .D68 2019 (print) | DDC 372.7/2–dc23
LC record available at https://lccn.loc.gov/2018055991

The National Council of Teachers of Mathematics advocates for high-quality mathematics teaching and learning for each and every student.

Printed in the United States of America

Contents

More4U

Foreword

Teaching mathematics in prekindergarten–grade 12 requires knowledge of mathematical content and developmentally appropriate pedagogical knowledge to provide students with experiences that help them learn mathematics with understanding, while they reason about and make sense of the ideas that they encounter.

In 2010 the National Council of Teachers of Mathematics (NCTM) published the first book in the Essential Understanding Series, focusing on topics that are critical to the mathematical development of students but often difficult to teach. Written to deepen teachers' understanding of key mathematical ideas and to examine those ideas in multiple ways, the Essential Understanding Series was designed to fill in gaps and extend teachers' understanding by providing a detailed survey of the big ideas and the essential understandings related to particular topics in mathematics.

The Putting Essential Understanding into Practice Series builds on the Essential Understanding Series by extending the focus to classroom practice. These books center on the pedagogical knowledge that teachers must have to help students master the big ideas and essential understandings at developmentally appropriate levels.

To help students develop deeper understanding, teachers must have skills that go beyond knowledge of content. The authors demonstrate that for teachers–

- understanding student misconceptions is critical and helps in planning instruction;
- knowing the mathematical content is not enough–understanding student learning and knowing different ways of teaching a topic are indispensable;
- constructing a task is important because the way in which a task is constructed can aid in mediating or negotiating student misconceptions by providing opportunities to identify those misconceptions and determine how to address them.

Through detailed analysis of samples of student work, emphasis on the need to understand student thinking, suggestions for follow-up tasks with the potential to move students forward, and ideas for assessment, the Putting Essential Understanding into Practice Series demonstrates best practice for developing students' understanding of mathematics.

The ideas and understandings that the Putting Essential Understanding into Practice Series highlights for student mastery are also embodied in the Common Core State Standards for Mathematics, and connections with these standards are noted throughout each book.

On behalf of the Board of Directors of NCTM, I offer sincere thanks to everyone who has helped to make this new series possible. Special thanks go to Barbara J. Dougherty for her leadership as series editor and to all the authors for their work on the Putting Essential Understanding into Practice Series. I join the project team in welcoming you to this special series and extending best wishes for your ongoing enjoyment—and for the continuing benefits for you and your students—as you explore Putting Essential Understanding into Practice!

Linda M. Gojak
President, 2012–2014
National Council of Teachers of Mathematics

Preface

The Putting Essential Understanding into Practice Series explores the teaching of mathematics topics in prekindergarten–grade 2 that are difficult to learn and to teach. Each volume in this series focuses on specific content from one volume in NCTM's Essential Understanding Series and links it to ways in which those ideas can be taught successfully in the classroom.

Thus, this series builds on the earlier series, which aimed to present the mathematics that teachers need to know and understand well to teach challenging topics successfully to their students. Each of the earlier books identified and examined the "big ideas" related to the topic, as well as the "essential understandings"—the associated smaller, and often more concrete, concepts that compose each big idea.

Taking the next step, the Putting Essential Understanding into Practice Series shifts the focus to the specialized pedagogical knowledge that teachers need to teach those big ideas and essential understandings effectively in their classrooms. The Introduction to each volume details the nature of the complex, substantive knowledge that is the focus of these books—*pedagogical content knowledge*. For the topics explored in these books, this knowledge is both student centered and focused on teaching mathematics through problem solving.

Each book then puts big ideas and essential understandings related to the topic under a high-powered teaching lens, showing in fine detail how they might be presented, developed, and assessed in the classroom. Specific tasks, classroom vignettes, and samples of student work serve to illustrate possible ways of introducing students to the ideas in ways that will enable students not only to make sense of them now but also to build on them in the future. Items for readers' reflection appear throughout and offer teachers additional opportunities for professional development.

The final chapter of each book looks at earlier and later instruction on the topic. A look back highlights effective teaching that lays the earlier foundations that students are expected to bring to the current grades, where they solidify and build on previous learning. A look ahead reveals how high-quality teaching can expand students' understanding when they move to more advanced levels.

Each volume in the Putting Essential Understanding into Practice Series also includes three appendixes. The appendixes list the big ideas and essential understandings related to the topic, detail resources for teachers, and present the tasks discussed in the book. These materials are available to readers both in the book and online at www.nctm.org/more4u, and are intended to extend and enrich readers' experiences and possibilities for using the book. Readers can gain online access to these materials by going to the More4U website and entering the code that appears on the book's title page. They can then print out these materials for personal or classroom use.

Because the topics chosen for both the earlier Essential Understanding Series and this successor series represent areas of mathematics that are widely regarded as challenging to teach and to learn, we believe that these books fill a tangible need for teachers. We hope that as you move through the tasks and consider the associated classroom implementations, you will find a variety of ideas to support your teaching and your students' learning.

Acknowledgments

The authors would like to thank Elisa Jannes (Principal) and the teachers of the Sigsbee Charter School in Key West, Florida, for collaborating with us on this volume. We would also like to thank the many kindergarten and first- and second-grade students who shared their thinking about mathematics problems and their enthusiasm for trying mathematical tasks. Thanks also is extended to Robert Ronau who created the figures for this book. We appreciate the support and knowledge of Anita Draper and Maryanne Bannon in their editing and fine eye for details.

Introduction

Shulman (1986, 1987) identified seven knowledge bases that influence teaching:

1. Content knowledge
2. General pedagogical knowledge
3. Curriculum knowledge
4. Knowledge of learners and their characteristics
5. Knowledge of educational contexts
6. Knowledge of educational ends, purposes, and values
7. Pedagogical content knowledge

The specialized content knowledge that you use to transform your understanding of mathematics content into ways of teaching is what Shulman identified as item 7 on this list—*pedagogical content knowledge* (PCK; Shulman 1986). PCK is the knowledge that is the focus of this book—and all the volumes in the Putting Essential Understanding into Practice Series.

Pedagogical Content Knowledge

In mathematics teaching, pedagogical content knowledge includes at least four indispensable components:

1. Knowledge of curriculum for mathematics
2. Knowledge of assessments for mathematics
3. Knowledge of instructional strategies for mathematics
4. Knowledge of student understanding of mathematics (Magnusson, Krajcik, and Borko 1999)

These four components are linked in significant ways to the content that you teach.

Even though it is important for you to consider how to structure lessons, deciding what group and class management techniques you will use, how you will allocate time, and what will be the general flow of the lesson, Shulman (1986) noted that it is even more important to consider *what* is taught and the *way* in which it is taught.

Every day, you make at least five essential decisions as you determine–

1. which explanations to offer (or not);
2. which representations of the mathematics to use;
3. what types of questions to ask;
4. what depth to expect in responses from students to the questions posed; and
5. how to deal with students' misunderstandings when they become evident in their responses.

Your pedagogical content knowledge is the unique blending of your content expertise and your skill in pedagogy to create a knowledge base that allows you to make robust instructional decisions. Shulman (1986, p. 9) defined pedagogical content knowledge as "a second kind of content knowledge . . . which goes beyond knowledge of the subject matter per se to the dimension of subject matter knowledge *for teaching*." He explained further:

> Pedagogical content knowledge also includes an understanding of what makes the learning of specific topics easy or difficult: the conceptions and preconceptions that students of different ages and backgrounds bring with them to the learning of those most frequently taught topics and lessons. (p. 9)

If you consider the five decision areas identified at the top of the page, you will note that each of these requires knowledge of the mathematical content and the associated pedagogy. For example, teaching students about number and numeration requires that you understand the centrality of the unit and place value concepts, and then determine contextual situations that best embody these ideas to present to your students. Your knowledge of number and numeration as related to quantities and relationships can help you craft tasks and questions that provide counterexamples and ways to guide your students in seeing connections across multiple number systems. As you establish the content, complete with learning goals, you then need to consider how to move your students from their initial understandings to deeper ones, building rich connections along the way.

The instructional sequence that you design to meet student learning goals has to take into consideration the misconceptions and misunderstandings that you might expect to encounter (along with the strategies that you expect to use to negotiate them), your expectation of the level of difficulty of the topic for your students, the progression of experiences in which your students will engage, appropriate collections of representations for the content, and relationships between and among the concept of unit, counting structures, prenumeric comparisons, place value understanding, and other topics.

Model of Teacher Knowledge

Grossman (1990) extended Shulman's ideas to create a model of teacher knowledge with four domains (see fig. 0.1):

1. Subject-matter knowledge
2. General pedagogical knowledge
3. Pedagogical content knowledge
4. Knowledge of context

Subject-matter knowledge includes mathematical facts, concepts, rules, and relationships among concepts. Your understanding of the mathematics affects the way in which you teach the content—the ideas that you emphasize, the ones that you do not, particular algorithms that you use, and so on (Hill, Rowan, and Ball 2005).

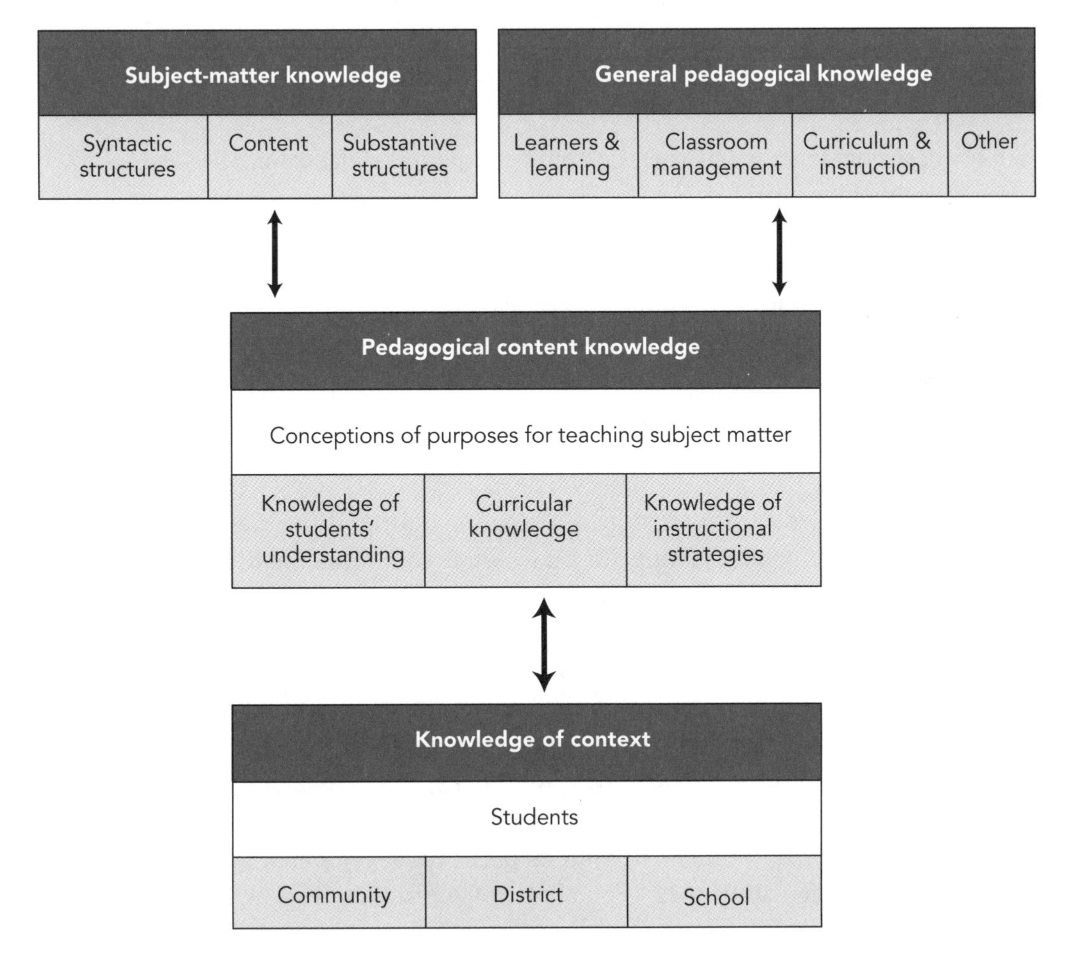

Fig. 0.1. Grossman's (1990, p. 5) model of teacher knowledge

Your pedagogical knowledge relates to the general knowledge, beliefs, and skills that you possess about instructional practices. These components include specific instructional strategies that you use, the amount of wait time that you allow for students' responses to questions or tasks, classroom management techniques that you use for setting expectations and organizing students, and your grouping techniques, which might include having your students work individually or cooperatively or collaboratively, in groups or pairs. As Grossman's model indicates, your understanding and interpretation of the environment of your school, district, and community can also have an impact on the way in which you teach a topic.

Note that pedagogical content knowledge has four aspects, or components, in Grossman's (1990) model:

1. Conceptions of purposes for teaching
2. Knowledge of students' understanding
3. Knowledge of curriculum
4. Knowledge of instructional strategies

Each of these components has specific connections to the classroom. It is useful to consider each one in turn.

First, when you think about the goals that you want to establish for your instruction, you are focusing on your conceptions of the purposes for teaching. This set of purposes is a broad category but an important one because the goals that you establish will define learning outcomes for your students. These conceptions influence the other three components of pedagogical content knowledge. Hence, they appropriately occupy their overarching position in the model.

Second, your knowledge of your students' understanding of the mathematics content is central to good teaching. To know what your students understand, you must focus on both their conceptions and their misconceptions. As teachers, we all recognize that students develop naïve understandings that may or may not be immediately evident to us in their work or discourse. These understandings can become deep-rooted misconceptions that are not simply errors that students make. Misconceptions may include incorrect generalizations that students have developed, such as thinking that in the number 22, the digit 2 in the tens place has the same value as the digit 2 in the ones place. These generalizations may even be predictable notions that students exhibit as part of a developmental trajectory, such as thinking that when adding 22 + 34, they are adding 2 + 3 rather than the actual value of 20 + 30.

Part of your responsibility as a teacher is to present tasks or to ask questions that can bring misconceptions to the forefront. Once you become aware of misconceptions in students' thinking, you then have to determine the next instructional steps. The mathematical ideas presented in this volume focus on common misconceptions that students form in relation to a specific topic—number and numeration. This book shows how the type of task selected and the sequencing of carefully developed questions can bring the misconceptions to light, as well as how particular teachers took the next instructional steps to challenge the students' misconceptions.

Third, curricular knowledge for mathematics includes multiple areas. Your teaching may be guided by a set of standards such as the Common Core State Standards for Mathematics (CCSSM; National Governors' Association Center for Best Practices and Council of Chief State School Officers 2010) or other provincial, state, or local standards. You may in fact use these standards as the learning outcomes for your students. Your textbook is another source that may influence your instruction. With any textbook also comes a particular philosophical view of mathematics, mathematics teaching, and student learning. Your awareness and understanding of the curricular perspectives related to the choice of standards and the selection of a textbook can help to determine how you actually enact your curriculum. Moreover, your district or school may have a pacing guide that influences your delivery of the curriculum. In this book, we can focus only on the alignment of the topics presented with broader curricular perspectives, such as CCSSM. However, your own understanding of and expertise with your other curricular resources, coupled with the parameters defined by the expected student outcomes from standards documents, can provide the specificity that you need for your classroom.

In addition to your day-to-day instructional decisions, you make daily decisions about which tasks from curricular materials that you can use without adaptation, which tasks you will need to adapt, and which tasks you will need to create on your own. Once you select or develop meaningful, high-quality tasks and use them in your mathematics lesson, you have launched what Yinger (1988) called "a three-way conversation between teacher, student, and problem" (p. 86). This process is not simple—it is complex because how students respond to the problem or task is directly linked to your next instructional move. That means that you have to plan multiple instructional paths to choose among as students respond to those tasks.

Knowledge of the curriculum goes beyond the curricular materials that you use. You also consider the mathematical knowledge that students bring with them from prekindergarten and what they should learn by the end of grade 2. The way in which you teach a foundational concept or skill has an impact on the way in which students will interact with and learn later related content. For example, the types of representations that you include in your introduction of number and numeration topics are the ones that your students will use to evaluate other representations and ideas in later grades.

Fourth, knowledge of instructional strategies is essential to pedagogical content knowledge. Having a wide array of instructional strategies for teaching mathematics is central to effective teaching and learning. Instructional strategies, along with knowledge of the curriculum, may include the selection of mathematical tasks, together with the way in which those tasks will be enacted in the classroom. Instructional strategies may also include the way in which the mathematical content will be structured for students. You may have very specific ways of thinking about how you will structure your presentation of a mathematical idea–not only how you will sequence the introduction and development of the idea, but also how you will present that idea to your students. Which examples should you select, and which questions should you ask? What representations should you use? Your knowledge of instructional strategies, coupled with your knowledge of your curriculum, permits you to align the selected mathematical tasks closely with the way in which your students perform those tasks in your classroom.

The instructional approach in this volume combines a student-centered perspective with an approach to mathematics through problem solving. A student-centered approach is characterized by a shared focus on student and teacher conversations, including interactions among students. Students who learn through such an approach are active in the learning process and develop ways of evaluating their own work and one another's in concert with the teacher's evaluation.

Teaching through problem solving makes tasks or problems the core of mathematics teaching and learning. The introduction to a new topic consists of a task that students work through, drawing on their previous knowledge while connecting it with new ideas. After students have explored the introductory task (or tasks), their consideration of solution methods, the uniqueness or multiplicity of solutions, and extensions of the task create rich opportunities for discussion and the development of specific mathematical concepts and skills.

By combining the two approaches, teachers create a dynamic, interactive, and engaging classroom environment for their students. This type of environment promotes the ability of students to demonstrate CCSSM's Standards for Mathematical Practice while learning the mathematics at a deep level.

The chapters that follow will show that instructional sequences embed all the characteristics of knowledge of instructional strategies that Grossman (1990) identifies. One component that is not explicit in Grossman's model but is included in a model developed by Magnusson, Krajcik, and Borko (1999) is the knowledge of assessment. Your knowledge of assessment in mathematics plays an important role in guiding your instructional decision-making process.

There are different types of assessments, each of which can influence the evidence that you collect as well as your view of what students know (or don't know) and how they know what they do. Your interpretation of what students know is also related to your view of what constitutes "knowing" in mathematics. As you examine the tasks, classroom vignettes, and samples of student work in this volume, you will notice that teacher questioning permits formative assessment that supplies information that spans both conceptual and procedural aspects of understanding. Formative assessment, as this book uses the term, refers to an appraisal that occurs during an instructional segment, with the aim of adjusting instruction to meet the needs of students more effectively (Popham 2006). Formative assessment does not always require a paper-and-pencil product but may include questions that you ask or tasks that students complete during class.

The information that you gain from student responses can provide you with feedback that guides the instructional flow, while giving you a sense of how deeply (or superficially) your students understand a particular idea—or whether they hold a misconception that is blocking their progress. As you monitor your students' development of rich understanding, you can continually compare their responses with your expectations and then adapt your instructional plans to accommodate their current levels of development. Wiliam (2007, p. 1054) described this interaction between teacher expectations and student performance in the following way:

It is therefore about assessment functioning as a bridge between teaching and learning, helping teachers collect evidence about student achievement in order to adjust instruction to better meet student learning needs, in real time.

Wiliam notes that for teachers to get the best information about student understandings, they have to know how to facilitate substantive class discussions, choose tasks that include opportunities for students to demonstrate their learning, and employ robust and effective questioning strategies. From these strategies, you must then interpret student responses and scaffold their learning to help them progress to more complex ideas.

Characteristics of Tasks

The type of task that is presented to students is very important. Tasks that focus only on procedural aspects may not help students learn a mathematical idea deeply. Superficial learning may result in students forgetting easily, requiring reteaching, and potentially affecting how they understand mathematical ideas that they encounter in the future. Thus, the tasks selected for inclusion in this volume emphasize deep learning of significant mathematical ideas. These rich, "high-quality" tasks have the power to create a foundation for more sophisticated ideas and support an understanding that goes beyond "how" to "why." Figure 0.2 identifies the characteristics of a high-quality task.

As you move through this volume, you will notice that it sequences tasks for each mathematical idea so that they provide a cohesive and connected approach to the identified concept. The tasks build on one another to ensure that each student's thinking becomes increasingly sophisticated, progressing from a novice's view of the content to a perspective that is closer to that of an expert. We hope that you will find the tasks useful in your own classes.

A high-quality task has the following characteristics:
Aligns with relevant mathematics content standard(s)
Encourages the use of multiple representations
Provides opportunities for students to develop and demonstrate the mathematical practices
Involves students in an inquiry-oriented or exploratory approach
Allows entry to the mathematics at a low level (all students can begin the task) but also has a high ceiling (some students can extend the activity to higher-level activities)
Connects previous knowledge to new learning
Allows for multiple solution approaches and strategies
Engages students in explaining the meaning of the result
Includes a relevant and interesting context

Fig. 0.2. Characteristics of a high-quality task

Types of Questions

The questions that you pose to your students in conjunction with a high-quality task may at times cause them to confront ideas that are at variance with or directly contradictory to their own beliefs. The state of mind that students then find themselves in is called *cognitive dissonance,* which is not a comfortable state for students—or, on occasion, for the teacher. The tasks in this book are structured in a way that forces students to deal with two conflicting ideas. However, it is through the process of negotiating the contradictions that students come to know the content much more deeply. How the teacher handles this negotiation determines student learning.

You can pose three types of questions to support your students' process of working with and sorting out conflicting ideas. These questions are characterized by their potential to encourage reversibility, flexibility, and generalization in students' thinking (Dougherty 2001). All three types of questions require more than a one-word or one-number answer. Reversibility questions are those that have the capacity to change the direction of students' thinking. They often give students the solution and require them to create the corresponding problem. A flexibility question can be one of two types: it can ask students to solve a problem in more than one way, or it can ask them to compare and contrast two or more problems or determine the relationship between or among concepts and skills. Generalization questions also come in two types: they ask students to look at multiple examples or cases and find a pattern or make observations, or they ask them to create a specific example of a rule, conjecture, or pattern. Figure 0.3 provides examples of reversibility, flexibility, and generalization questions related to number and numeration.

Type of question	Example
Reversibility question	Corey used exactly five base-ten blocks to represent a number. What number could Corey have represented?
Flexibility question	Using base-ten blocks, how can you represent 27? Find three more ways to represent 27.
Flexibility question	Janese represented 23 using two 10s and three 1s with base-ten blocks. What number could she represent if she adds one more block to her presentation?
Generalization question	Yosh counted out loud by 10s. What pattern(s) might Yosh notice about the numbers?
Generalization question	Jenna noticed that when she counted by 10s to 50, it was faster than counting by 5s. Why would Jenna say that? Explain.

Fig. 0.3. Examples of reversibility, flexibility, and generalization questions

Conclusion

The Introduction has provided a brief overview of the nature of–and necessity for–pedagogical content knowledge. This knowledge, which you use in your classroom every day, is the indispensable medium through which you transmit your understanding of the big ideas of the mathematics to your students. It determines your selection of appropriate, high-quality tasks and enables you to ask the types of questions that will not only move your students forward in their understanding but also allow you to determine the depth of that understanding.

The chapters that follow describe important ideas related to learners, curricular goals, instructional strategies, and assessment that can assist you in transforming your students' knowledge into formal mathematical ideas related to number and numeration. These chapters provide specific examples of mathematical tasks and student thinking for you to analyze to develop your pedagogical content knowledge for teaching number and numeration in prekindergarten–grade 2 or to give you ideas to help other colleagues develop this knowledge. You will also see how to bring together and interweave your knowledge of learners, curriculum, instructional strategies, and assessment to support your students in grasping the big ideas and essential understandings and using them to build more sophisticated knowledge.

Students entering prekindergarten have already had some experiences that affect their initial understanding of number and numeration. Furthermore, they have developed some ideas about these topics at earlier levels. Students in the first years of school frequently demonstrate understanding of mathematical ideas related to number and numeration in a particular context or in connection with a specific physical material, picture, or drawing. Yet, in other situations, these same students do not demonstrate that same understanding. As their teacher, you must understand the ideas that they have developed about number and numeration in their prior experiences so you can extend this knowledge and see whether or how it differs from the formal mathematical knowledge that they need to be successful in reasoning with or applying counting or other numerical concepts and skills. You have the important responsibility of assessing their current knowledge related to the big ideas of number and numeration as well as their understanding of various representations of these concepts and their power and limitations. Your understanding will facilitate and reinforce your instructional decisions. Teaching the big mathematical ideas and helping students develop essential understandings related to number and numeration is obviously a very challenging and complex task.

into practice

Chapter 1
Prenumeric Ideas

Big Idea 1

Number is an extension of more basic ideas about relationships between quantities.

Essential Understanding 1*a*

Quantities can be compared without assigning numerical values to them.

Essential Understanding 1*b*

Physical objects are not in themselves quantities. All quantitative comparisons involve selecting particular attributes of objects or materials to compare.

Essential Understanding 1*c*

The relation between one quantity and another quantity can be an equality or inequality relation.

Essential Understanding 1*d*

Two important properties of equality and order relations are conservation and transitivity.

In *Developing Essential Understanding of Number and Numeration for Teaching Mathematics in Prekindergarten–Grade 2* (2010), Dougherty and colleagues noted that understanding the concept of number is much more than learning how to rote count or how to count a discrete number of objects in a set. The concept of number must be first understood by recognizing that counting provides a measure of a quantity or a set of objects. Counting is based on the idea that some identified unit is used to determine how many of that unit is present in the quantity or set being counted (Dougherty and Venenciano 2007).

However, this process of counting assumes that there is an understanding of quantities and relationships that can be present within and across quantities. This understanding is the conceptual aspect of interpreting number and forms the foundation of number development from its inception before prekindergarten and beyond.

Foundational Ideas about Number and Numeration

Using prenumeric comparisons to understand quantities and relationships

The foundational aspects of number begin by first identifying what attributes of objects can be compared and, in essence, measured. These attributes may be the length, area, volume, or mass of an object (Dougherty and Venenciano 2007). By directly comparing the attributes of objects, students can determine equality or inequality. The explicit identification of attributes that can be measured and compared helps students to differentiate among the attributes that are more difficult to compare and measure and may not be mathematical in nature at all, such as color. If you've never thought deeply how young children develop ideas of equivalence, Reflect 1.1 invites you to do so now.

Reflect 1.1

Look at figure 1.1. What ideas about equivalence can come from student observations of the direct comparison?

How might the attributes of volume, area, or mass be used to demonstrate equivalence relationships?

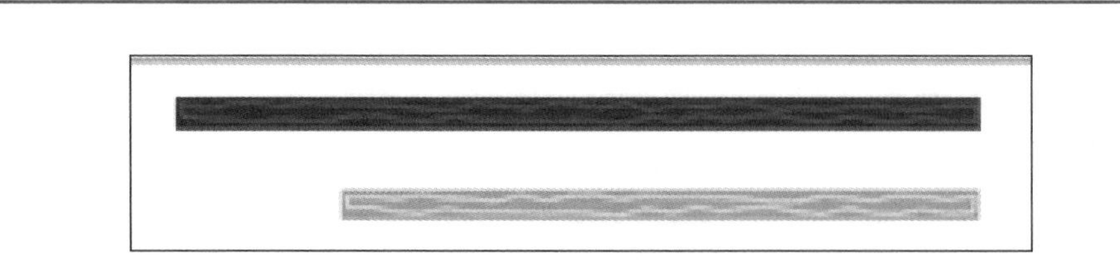

Fig. 1.1. Direct comparison of the length of two straws (Dougherty et al. 2010, p. 10)

For example, in figure 1.1, the lengths of two straws are compared by aligning one end of each straw with the other. By placing them in this manner, students can see that the length of one straw is greater than the length of the other straw OR the length of one straw is less than the length of the other straw.

There are many benefits in having students compare lengths, areas, volumes, and masses as part of the foundational development of number. For young children to determine which number is greater (or lesser or equal) when presented with 3 and 5, they need a mental picture of what these amounts might look like when two quantities are not equal. The mental picture of the quantitative relationship is developed through prior consistent use of physical materials that can be manipulated in multiple ways to model the relationship and then make an interpretation of what is being modeled. This picture is called a *mental residue* (Dougherty 2008), and it provides a way for students to access these comparisons more abstractly when they might only be represented with numerical quantities.

These initial experiences in number can also provide a way to illustrate the importance of the precision of language. Words like *bigger, littler,* and *huger* are often used by young children but are not necessarily indicative of the attribute or quality that was being compared.

Comparing unlike objects

Twenty-four kindergarteners and thirty first-graders were shown the two objects displayed in figure 1.2 and asked to decide which was larger. Consider in Reflect 1.2 how children might react when encountering an "apple to oranges" comparison situation.

Fig. 1.2. Comparison of size of two objects

Reflect 1.2

Consider the comparison task given above. How might young children respond to the question?

On what attribute(s) might they focus? Why?

All of the kindergartners and 86.6 percent (26 out of 30) of the first graders responded that the blue paper was "bigger." When asked why it was larger or greater than the pen, students had a difficult time explaining why they chose it. Corey, a kindergartener, said, "Well, it just looks bigger 'cause the pen looks littler." Micah, a first grader, said, "Everybody can see it's bigger. Can't you see it?" Their responses are typical of all the students who selected the blue paper.

Conversely, the four first graders who did not select the blue paper had other ideas. June said, "Hmm, it's hard for me." When asked why, she said, "'Cause the pen is longer but the paper is bigger this way [*motioning to the width of the paper*]. So, I don't know." Naveah had a similar explanation: "I don't know what larger means when they [objects] are different. Maybe the pen? But maybe the paper?"

From their responses, you can see that the concepts of *greater than* or *less than* are relevant to the attribute of the quantities being compared and students are not sure how to use those attributes. When we ask questions such as, "Which one is larger?" students may be confused about what they are actually comparing.

Notions of equality

While the modeling of the relationships of greater than, less than, equal, and unequal are important, there are even more powerful ideas that stem from this component of prenumber development. Without using numbers, students can explore some properties of equality at a very early age. Consider the following task adapted from the Measure Up project (Dougherty and Venenciano 2007). Before presenting the task to your students, take time to consider Reflect 1.3. Even without a formal understanding of volume, pre-K and kindergarten children have a sense of equal and unequal amounts and how to make them "the same."

Task: Two Bottles of Water

Sammi had two bottles of water. She wanted to have the same amount of water in both bottles, but she cannot fill the bottles to the top. What could she do?

Figure 1.3. Sammi's water bottle comparison

Reflect 1.3

Consider the task presented above. What properties of equality might students use? What background knowledge might they have that could support ideas they could share?

First-grade students brainstormed ideas about what Sammi could do. To facilitate their discussion, the teacher labeled the water on the left as volume *C* and the water on the right as volume *M*. She identified the volumes of water in this way to help students better communicate their proposed actions and to provide a way to document what they said and the result of their actions. Note that it is the *volume* of water in the bottle that is labeled, and not the bottle itself.

Larry proposed that Sammi could pour other water into volume C until the water level was the same as volume M. Tori thought that another way would be to pour water out from volume M until the height of the water was the same as volume C. The teacher asked how the amount of water added to volume C and the amount of water taken out of volume M were related. The students all agreed that the volume of water removed and the volume of water added would be the same. The teacher then identified the volume removed or added as volume H. She called this quantity of water the *difference*.

This discussion resulted in the students and teacher creating the following equations:

$$C + H = M$$
$$M - H = C$$

Including the symbolic representation (equations) gives students a meaningful way to document the process they used and attaches a language to the physical action.

This type of task is powerful because it encourages students to think about relationships that cannot be easily modeled and understood at such an early age. The big idea that emerges here for children is that when two quantities are unequal, they can be made equal by adding or subtracting the same amount from the respective quantity. This idea is an important underpinning for future number work because it illustrates *difference* in a meaningful way.

Further tasks should focus on maintaining equality, given two quantities that are equal. For example, if two masses on a balance scale are equal, what can be done to add or take away mass on both sides to maintain equality? Students notice that if the same amount is added or taken away to both sides of the balance scale, the masses will remain equal. This important concept about number and equality will support student understanding as children move forward in their development of algebraic thinking.

Also, in this prenumeric stage of development, students can begin to apply the reflexive, symmetric, and transitive properties of equality, which are rarely discussed in the early grades. These properties are directly related to the first-grade standard in the Common Core State Standards for Mathematics (CCSSM; National Governors Association Center for Best Practices and Council of Chief State School Officers 2010).

Common Core State Standards for Mathematics

Related to the Big Idea and Essential Understandings for Chapter 1

Grade 1 (1.OA.7)

7. Understand the meaning of the equal sign, and determine if equations involving addition and subtraction are true or false. *For example, which of the following equations are true and which are false?* $6 = 6$, $7 = 8 - 1$, $5 + 2 = 2 + 5$, $4 + 1 = 5 + 2$.

(National Governors Association Center for Best Practices and Council of Chief State School Officers [NGA Center and CCSSO] 2010.)

This standard is often interpreted as students being able to demonstrate that $3 + 2 = 5$ but the meaning of the equal sign is much more than a symbol in an equation. In the pre-numeric stage, students recognize that the equal sign indicates that two quantities represent the same amount. The equal sign is not a signal that the "answer" comes next as it is often interpreted.

The three properties of equality (reflexive, symmetric, and transitive) can all be modeled by using physical materials. For example, it is clear that an area, say the area of a tabletop, must be equal to itself. Hence, if the area of a tabletop is represented by area K, then we can symbolize that relationship as $K = K$ (reflexive property).

The symmetric property is easily modeled by having two congruent areas, area F and area P. Students can directly compare the two areas by laying one area on top of the other and seeing whether there are any overlaps. As students describe what they notice about the areas being the same, some will naturally say area F is equal to (the same as) area P while others will say that area P is equal to area F. This relationship should be recorded as both $F = P$ and $P = F$.

The transitive property of equality is initiated by having students compare, for example, three areas such as, area B, area L, and area W. By direct comparison, they determine that $L > B$. The teacher asks students to find out how area B compares to area W without moving the region that represents area B. Students suggest that instead, they can compare area L to area W, and upon doing that, they discover that $L = W$. Because those two areas are equal, students can now surmise that $B < W$ (or $W > B$).

The transitive property of equality is more complex than the reflexive and symmetric properties, but it, too, is accessible by very young students when physical materials are used. Number development without using numbers seems contradictory, but this approach provides opportunities for students to consider the quantitative relationships that are aligned with significant number concepts and use them with greater understanding.

Summary: Learners, Curriculum, Instruction, and Assessment

To effectively teach the mathematical ideas presented in this chapter, teachers must have knowledge of the four components–learners, curriculum, instructional strategies, and assessment–presented in the Introduction. The following sections summarize some key ideas for each of these elements.

Knowledge of learners

Young students often do not have adequate opportunity to learn the foundational aspects of number, particularly those concepts associated with quantitative relationships. They have not yet gained the sophistication in their thinking to solely work with symbolic representations. And, the exclusive use of discrete sets of objects to count and compare quantities does not provide a context that can be generalized to other number systems, such as rational numbers. The use of physical objects combined with a symbolic recording of the relationships between the objects fits with young children's developmental level and affords them the opportunity to work with more complex mathematical ideas and truly focus on conceptual understanding.

Knowledge of curriculum

The understanding of the meaning of the equal sign and its related properties is critical in constructing a strong foundation for students' number knowledge. Building on experiences from pre-K and the early grades, students need to acquire the realization that the equal sign indicates that the two quantities being compared represent the same amount; otherwise, they consider the equal sign as indicating that the answer comes next. The explicitness of instruction that focuses student attention on quantitative reasoning provides more opportunities for students to truly see the meaning of this most important symbol. Additionally, the properties of equality (or inequality) lead to other number understandings, but these properties are often assumed and not overtly taught.

Knowledge of instructional strategies

The careful use of language is a critical consideration when helping students develop understanding of number. Careful and explicit identification of the attributes of objects being compared or measured focuses students' attention and allows them to construct more precise descriptions of the relationships they notice. This attention to units leads, even in the earlier grades, to substantive mathematical discussions where students can justify their reasoning and confront misconceptions.

Knowledge of assessment

Assessments, such as formative assessments generated by the tasks presented, should provide a window into students' thinking and understanding. Children can be asked to interpret physical representations and justify their thinking about a relationship between these representations. Conversely, students can be given a relationship and asked to provide a model that shows it. We can initially be vague in our description of a relationship, such as the book is bigger than the sheet of paper, and then ask students to find better ways of describing and conveying the relationship. Tasks should be varied to promote student flexibility in considering and interpreting quantitative relationships.

Conclusion

In order to help children develop a deep understanding of number, teachers must provide their students with opportunities to interpret and model relationships using physical objects along with symbolic representations. Focusing on precise language within the instructional tasks is particularly important in developing ways of communicating about relationships as well as supporting the structures represented in the models. The deep understanding of these concepts will form the basis for further work with number by developing the concept of a unit and how it supports the comparison of numerical quantities, the focus of chapter 2.

into practice

Chapter 2
The Concept of Unit

Big Idea 2
The selection of a unit makes it possible to use numbers in comparing quantities.

Essential Understanding 2*a*
Using numbers to describe relationships between or among quantities depends on identifying a unit.

Essential Understanding 2*b*
The size of a unit determines the number of times that it must be iterated to count or measure a quantity.

Getting to Understanding the Concept of *Unit*

When children start to think about using numbers to count quantities, what separates that process from their previous use of non-numerical descriptions (e.g., "a lot," "humoungous," and so on) is the formal arrival of the use of a unit. When a number is given to communicate a quantity, the child should recognize that we are attaching a value to a collection or a measurement based on a particular unit. That unit can start as a comparison of a measure of length or time, for example, "I am three!" or it can be the documentation of a number of objects, such as "I have two Lego figures."

During this developmental period, students must start to explicitly see the relationship between the count, or how many, and the unit used. Are we counting the number of mittens or are we counting pairs of mittens? By establishing a unit, "the numerical value we give essentially represents the multiplicative relation between our chosen unit and the total quantity" (Sophian 2007).

Unitizing is attaching a specific quantity to the unit of measure. The unit represents the piece, or the portion, that we are using as our counting unit. This idea will become even more important as students begin to think about multiplication, division, and fractions (Lamon 1993a; 1993b). For example, Lamon suggests considering the image of a case of soda as either 24 single unit cans, two 12-packs, or four packs of 6. As students become more practiced in identifying this relationship, they begin to recognize that when measuring with two different units, the same length or set will have a larger measure with a smaller unit and a smaller measure with a larger unit.

Challenges in making comparisons of quantities

When children first begin to compare two quantities, the tasks that have students considering small differences are much more challenging than the problems with large differences. So, for example, students have a harder time discerning whether there are more grapes when given a choice of seven or nine grapes versus the choice of seven or seventeen grapes (Feigenson, Dehaene, and Spelke 2004). These problems are called conflict/agreement problems (Sophian 2000). This struggle can also be an issue when children look at sets and compare larger amounts of smaller items with smaller amounts of large items. Reflect 2.1 raises questions on how children might approach such situations while grappling with these comparison relationships.

Let's look at this task, shown in figure 2.1, adapted from the work of Sophian.

Task: Feed the Puppet

Show students the following plates of cookies. Say Hungry Puppet wants to eat the most cookies. Which plate has the most cookies? Ask the student to circle their choice.

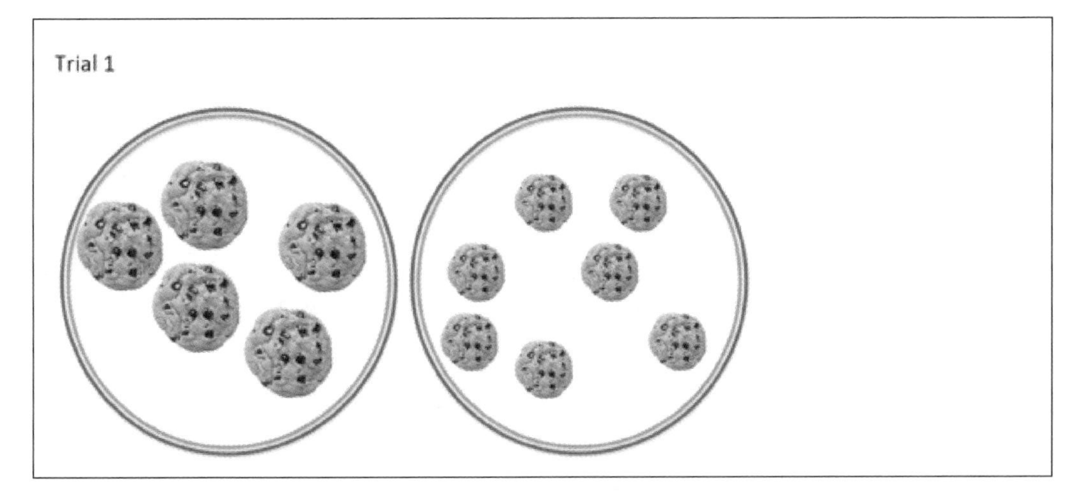

Fig. 2.1 Sample task from the Cookie Comparison problem

Reflect 2.1

How might students interpret the task?

What ideas might they use to think about which plate to choose?

How do these ideas relate to the concepts presented in chapter 1?

These tasks set a context of a hungry puppet who wants to eat as much as possible by choosing between alternative collections that differ in the size of the cookies as well as in the total number of cookies. In successive tasks students make comparisons with different units, and without any suggestion from the teacher, assign a value to identify which plate has more.

Select two plates (as in figs. 2.1–2.5) and ask the student, "Which plate would the hungry puppet want to select?" Presented below are a variety of responses from preschoolers and kindergarteners.

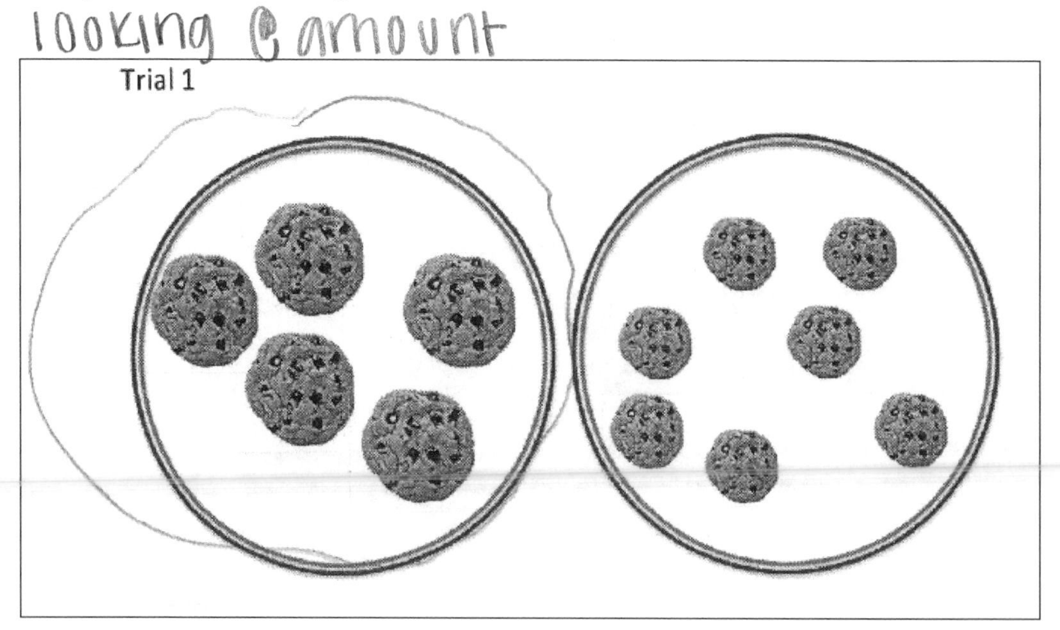

Fig. 2.2. Luca's solution to trial 1 of the Cookie Comparison problem

Here, Luca reasoned as he counted the five cookies on the plate, "Because he is super hungry and he wants the mostest cookies."

Luca circled the following choice using the opposite reasoning when he stated, "He likes the mostest cookies, five" without counting.

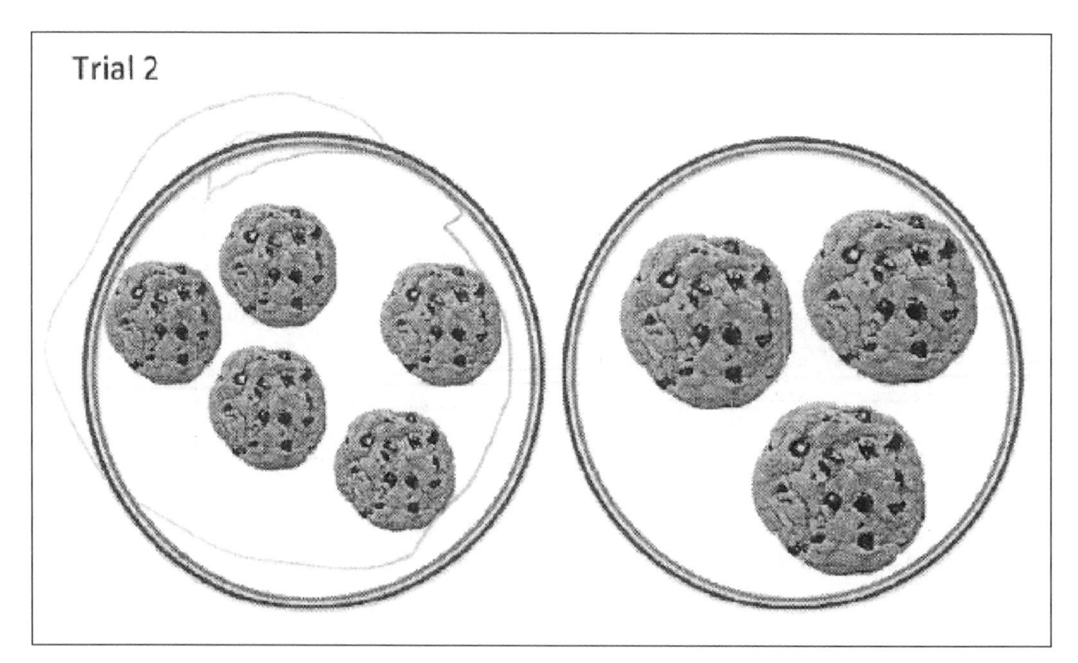

Fig. 2.3. Luca's response to trial 2 of the Cookie Comparison problem

Interestingly, on the final trial Luca shifted thinking again back to the largest cookie rather than the largest number when he stated, "It only has one; the other has seven."

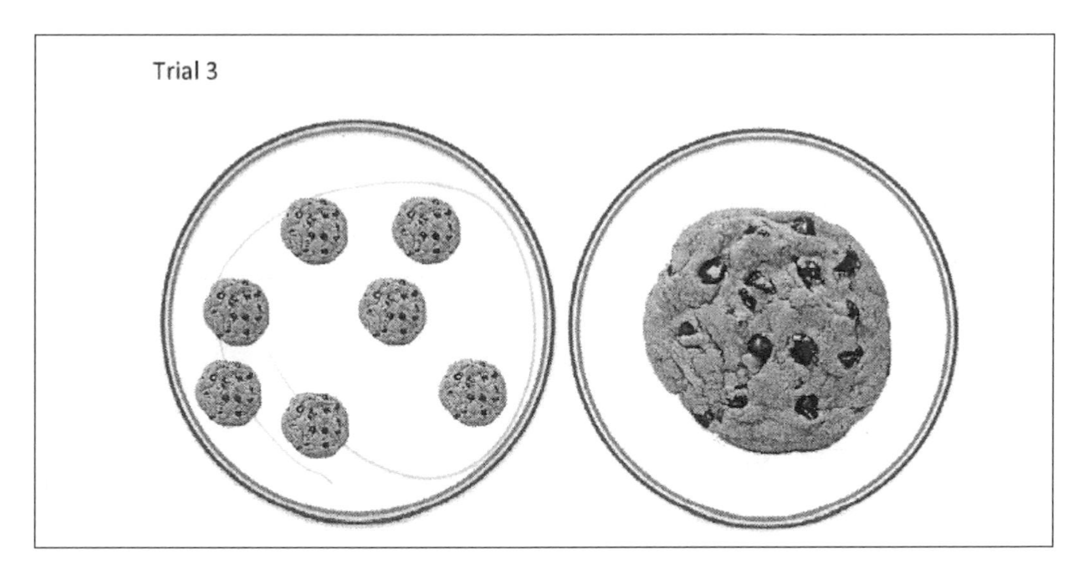

Fig. 2.4. Luca's response to trial 3 of the Cookie Comparison problem

Other students, such as Rani, had identifiable patterns in their decision-making.

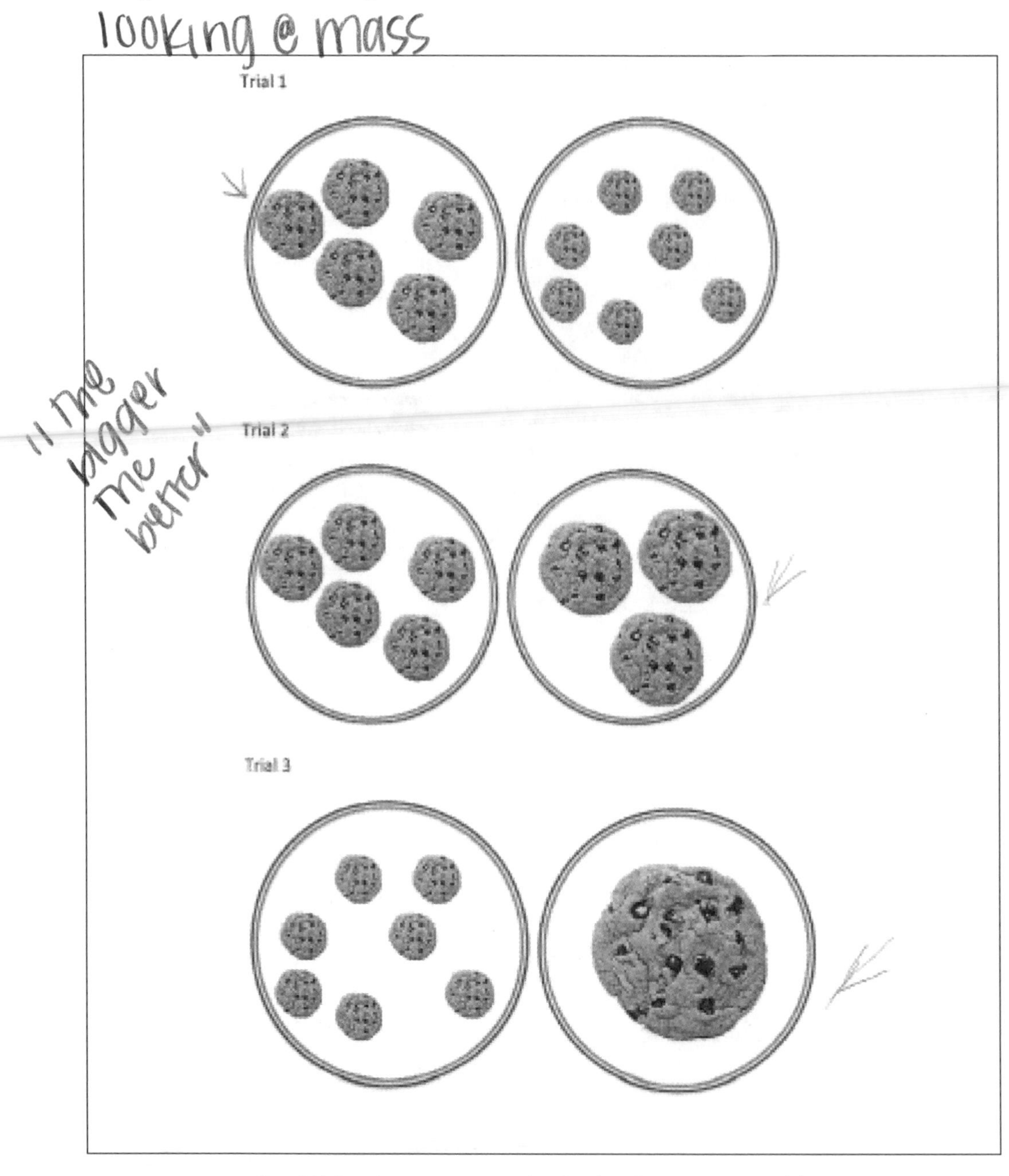

Fig. 2.5. Rani's responses to the Cookie Comparison problem

You can discern Rani's pattern as she consistently selects the "bigger meal" (the larger size cookie) saying the puppet is "very hungry."

In this next sample, Jamie unfailingly counted out the cookies in each trial, even recounting the plate of five in the second trial. Here are her choices:

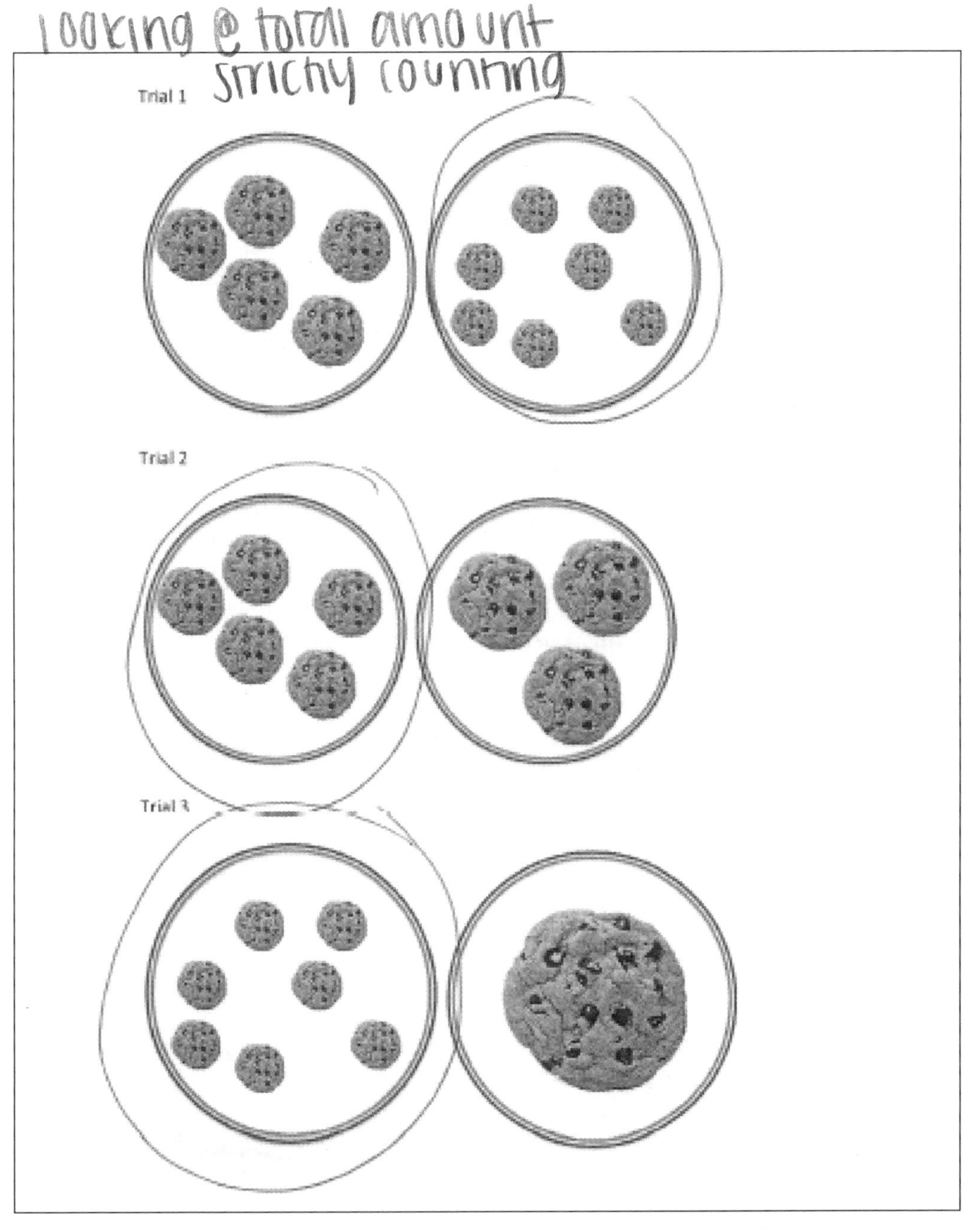

Fig. 2.6. Jamie's responses to the Cookie Comparison problem

As she touched each cookie she commented, "It's a lot. It has five and it has seven." Then she circled the five cookies in trial 2 and said, "'Cause it has five and this one has three." Finally, she still counted the last set and circled the seven cookies saying, "It had more. Lots."

Many children, even those who counted each plateful, were swayed by the one large cookie in the final trial, and rightfully so. Evaluating the single cookie by the measure of its area, it has more cookie than the seven on the other plate. Children described the single cookie option as "Lots," "Most biggest cookie," "Gigantest," and it "Fills the whole thing."

As you can see from the student work of these four- and five-year olds, this cookie comparison is not a straightforward task for that age group. The very youngest children tend to pick the plate with the biggest cookies even though there are more cookies on the other plate choices. This phenomenon was replaced as children got older; then they began to see the largest amount as related to counting the number of objects regardless of the size of the individual object.

Asking students to judge the magnitude of a set of objects or to describe the relationship between two quantities without first identifying the attribute and the unit being compared or measured can confuse them. In the case of the cookie task, students clearly used different ways in which to determine the "largeness" of the sets. Some used the area of the cookies while others used the cardinality of the number of cookies, even though the size of the cookies was not consistent.

Understanding how to use a defined unit to measure or to count a quantity is another building block in students' foundation of number. Often assumed, it is without a doubt one of the explorations that is not explicitly examined with students. It is, however, mentioned as a standard in grade 2 of the Common Core State Standards, but notice that the standard is in the Measurement and Data domain, and not specifically linked to number.

Common Core State Standards for Mathematics

Related to the Big Idea and Essential Understandings for Chapter 2

Grade 2 (2.MD.2.)

2. Measure the length of an object twice, using length units of different lengths for the two measurements; describe how the two measurements relate to the size of the unit chosen.

(National Governors Association Center for Best Practices and Council of Chief State School Officers [NGA Center and CCSSO] 2010.)

Let's look at a task that relates measurement to number. Consider the task where students are given two areas as shown in figure 2.7. Reflect 2.2 challenges you to anticipate how young students with no knowledge of the traditional ways to calculate area might approach the task.

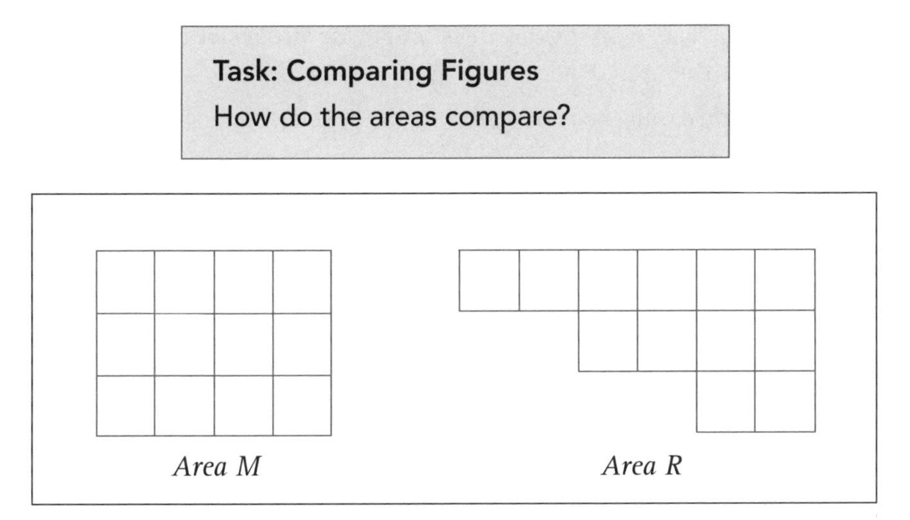

Fig. 2.7. Task given to first-grade students

Reflect 2.2

Given that first-grade students do not know how to compute area, how might they solve the problem?

What is the importance in defining a unit in the task?

Because the two areas were partitioned into small squares, some students naturally began by counting them, using those small squares as their unit. An interview with Ella describes some of the ideas that emerged from this task.

Ella: I wrote $M = R$ because there are 12 of the same unit in both of them.

Teacher: Are there other units you could have used?

Ella looks at the figures and outlines two of the squares.

Ella: I could use two of small ones to make a unit.

Teacher: If you use that new unit, do you think the areas of the two shapes will still be equal?

Ella: I'm don't know.

Ella iterates the two units on both areas.

Ella: They are! It takes six of these units!

Teacher: Hmm, that's interesting. It takes 12 of the smaller units to cover the squares, but only 6 of the new unit you used.

Ella: That's 'cause my new unit is bigger, so it doesn't take as many to make the area.

Jess: *(Overhearing the conversation)* Yeah, the bigger the unit, the lesser you use.

Ella: Yeah! And the smaller the unit, the more you use!

The conversation went further as students created more units and began comparing them and the counts that resulted. For example, they found that if the unit consisted of four of the smaller squares, the count would be 3. They thought they had found all the whole number values of squares that could be used to measure the areas when the teacher asked if there were any more units that could be used to compare the areas.

A pair of students shared with the class that they thought you could subdivide one of the squares into two smaller units, that is, in half. If you did that for each of the small squares, the area would be counted as 24 units. That led the class to further discuss the relationship between the size of the unit and the resulting count.

This example is focused on a geometric representation involving area. However, the central idea of the relationship between unit and the resulting count extends into more traditional tasks that focus on comparing numbers. For example, when students are asked to compare 8 and 15, it is assumed that the same unit was used to derive the counts of 8 and 15.

First graders in the Measure Up project (Dougherty 2008), which includes a focus on the importance of unit, were presented with the statement $3 < 8$. Below is an excerpt of their discussion.

Curt: Hmm, 3 is less than 8.

Teacher: Yes, is that a true statement?

Megan: But 3 might not be less than 8.

Teacher: When might that happen?

Megan: Well, you could have 3 really, really, really big units and 8 really, really, really small units.

Alisha: Yeah, then 3 would be greater than 8!

Davon: But, if this was on a number line, then 3 would be less than 8 'cause the units would all be the same.

This idea of considering the size of the unit when making comparisons can also be investigated by young children with the attribute of length. Explore the Which Snake is the Longer? task with your students (see fig. 2.8 and Appendix 3 and More4U for materials you can copy and cut out).

> **Task: Which Snake Is Longer?**
>
> Students are asked to compare two snakes visually and then by counting different-size units.

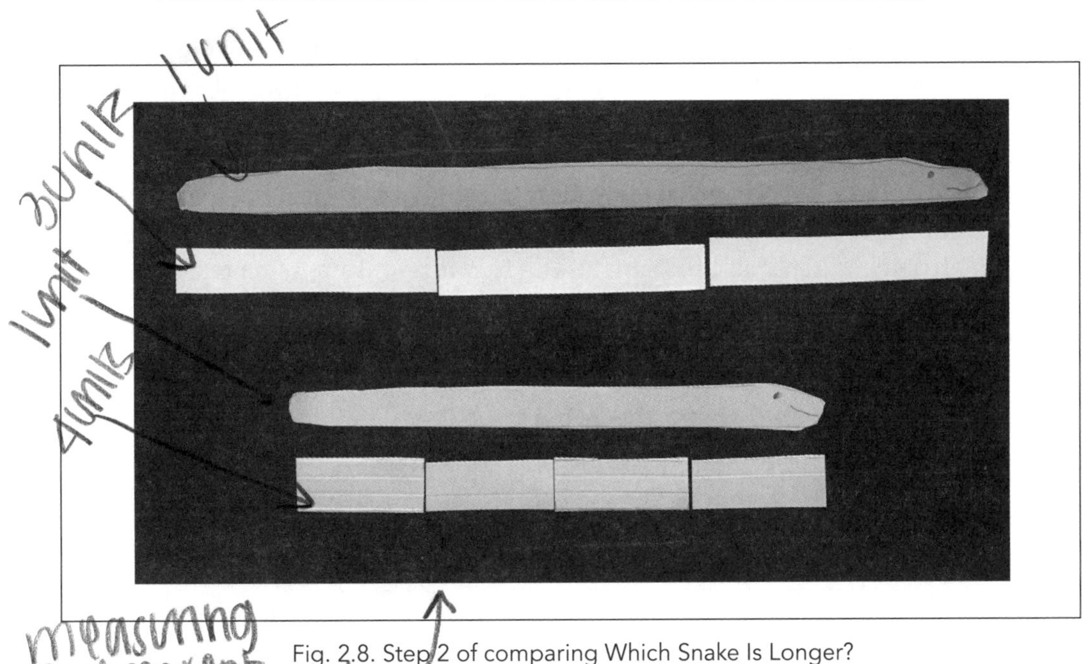

Fig. 2.8. Step 2 of comparing Which Snake Is Longer?

First students are asked to consider the lengths of the two snakes placed on the table and to decide which snake is longer. After they respond that the green snake is longer, take out the units and place them down in a row under each snake, counting as you go. Say, “I counted that the green snake is 3 units long (using the longer units) and the yellow snake is 4 units long (using the shorter units). Doesn’t that make the yellow snake longer?”

This is where the children are challenged by the cognitive dissonance that contrasts what their own observations suggest and what the evidence of the unit count of a measure produces. One student said, “Maybe I was too fast” after initially selecting the green snake and then hearing from the teacher that the yellow snake was 4 units and the green was 3. Then she said, “No. Because these are little pieces and the others are big. Pretend this is like a road. One little piece takes up littler space and the big piece takes up more space. The green snake is still longer.” Another child said, “Use the same pieces to measure. You can’t have big for the one snake and little for the other snake. Not fair.”

We cannot emphasize enough the importance of this foundational concept for young learners—that the defined unit impacts the measure or the count—and how it plays out for students as they progress in school.

Conservation and Counting

Counting plays a critical role in children's ability to compare two sets of objects (Saxe 1977; Sophian 1987; 1988; 1995). "It is precisely this use of counting that is most relevant to conservation problems" (Sophian 1995, p. 560) because children need to count to make comparisons between two sets (using one-to-one correspondence) before they can reach success on conservation tasks. Research from Sophian (1995) confirms this trajectory, suggesting that the identification of a unit and its corresponding attribute is applicable to these tasks. (This relationship, as represented in the Setting the Table for a Party task, is the subject of Reflect 2.3.)

Task: Setting the Table for a Party

Students are asked to compare the same number of spoons and forks after they are put in rows with different spacing.

A task (based on the classic Piagetian conservation task; 1952) using the scenario of a party and 8 plastic spoons and 8 plastic forks provides information about the sophistication of a child's understanding of identifying the unit and the associated attribute. The directions are as follows: Show the child 8 plastic spoons and 8 forks paired and placed as you see in figure 2.9.

Fig. 2.9. Fork and spoon task for conservation, part 1

Ask the child, "Do I have more spoons or more forks?"

Then ask the child to watch you as you move the spoons in full view of the student in the position shown in figure 2.10 in relation to the forks. Make sure you move the objects so that the student sees that you are using the same number of objects as in the previous question without adding more utensils.

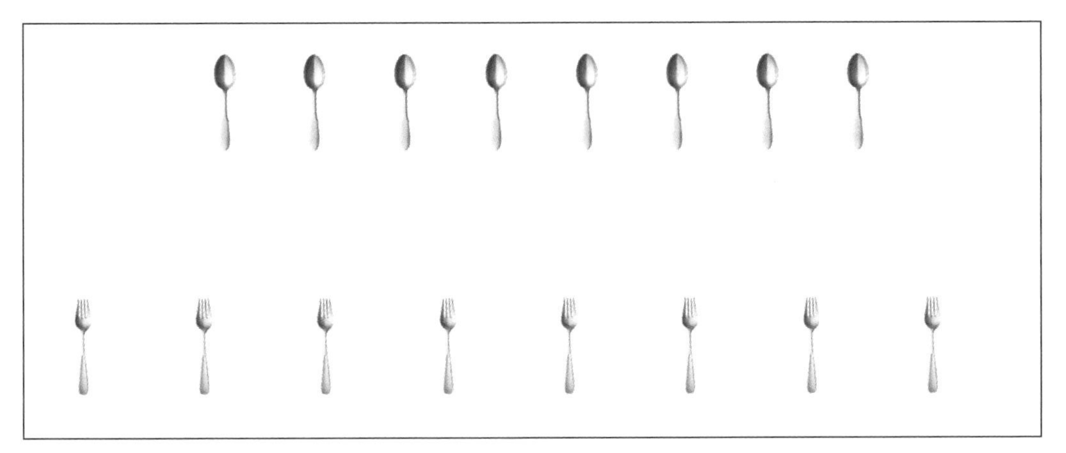

Fig. 2.10. Fork and spoon task, part 2

Then ask, "Do I have more spoons or more forks?"

Reflect 2.3

How does this conservation task align with the concept of unit?

How might a student's understanding of attributes and the associated unit impact their response to this task?

If the child states that there is an equal amount, test the strength of resolve behind the response by then saying, "It looks like there are more forks than spoons." This follow-up statement is to determine the child's level of confidence in the response as well as to set up the opportunity for the child to prove their answer or discuss their thinking.

The following expected levels of performance are based on the work of Piaget (1965):

- Level 1–The child says they don't know whether there are more spoons or forks.
- Level 2–The child says there are more forks.
- Level 3–The child says there are the same number of spoons and forks, but when you question that response by saying "There looks like there are more forks," the child relents and agrees that there are more forks.
- Level 4–The child says there are the same number of spoons and forks, and when questioned with the level 3 "doubt" question, the child proves that there are as many spoons as there are forks by matching them in a one-to-one correspondence or by counting.

When this task was given to a group of twelve preschoolers, ten suggested there were more forks than spoons. One child who had suggested there were equal amounts of spoons and forks switched to saying that there were more forks after being questioned about his choice. One child specifically described the relationship as "two more," counting the forks that extended beyond the spoons on either side of the length. Another child described the difference as "lots of forks." This outcome is not surprising as students at that age are influenced by variables other than counting, and the space the forks take up in the line of comparison is a compelling choice for which is larger. Thus, the attribute they are using is length while the question is actually focused on the number of forks or spoons, using a utensil as a unit. As a follow-up, the students were asked to count each collection, and in the case of all the efficient counters, these students changed their response to saying that there were the same number of forks and spoons.

This type of thinking also connects to the two core areas–Number and Relations–detailed in the National Research Council Early Childhood Committee Math Report (2009). The report makes the case that these core areas of Number and Relations connect to the appearance of a developmental trajectory that extends from grasping the numerical value of matching sets to comprehending the relationship between quantities in two nonmatching sets (Purpura and Lonigan 2013). The emphasis on understanding number and relationships in the NRC report encouraged the focus on these concepts rather than the rote-teaching methods that often did not support the learning of all children and were often described as "teaching without learning" (Fuson 2009, p. 344).

The case is made that as children mature, they begin to do better on the first type of conservation task shared above where they see that an amount of items is the same number as another set of items; then they progress to recognizing that the number of forks and spoons remains the same even when the items are moved to positions that take up a larger or a lesser area. The children learn to address the issue of the alteration of the space as well as the comparison of the quantities of two different types of items. This fork and spoon task aligns with the classic work of Elkind (1967) and others (Schaeffer, Eggleston, and Scott 1974; Sophian 1987; 1988) where the distinction between these conservation tasks persisted among children, although assessing the relationship between two quantities remains the same regardless of the movement of the items.

This look at conservation can be examined in another way, again by bringing the attribute of length into the decision-making process. In this example, two sets of peanuts, both aligned with a left justification (adapted from Sophian 1995, p. 561), are being compared. Confusion arises when the second row of peanuts has the same length as the first as shown in part 2 of the task.

Task: Rows of Peanuts, Part 1

Children compare the number of peanuts in two rows of unequal length. They are asked to select the greater amount

Kindergarten students are given the peanut task presented in figure 2.11 where there is an unequal number of objects in two rows of unequal length.

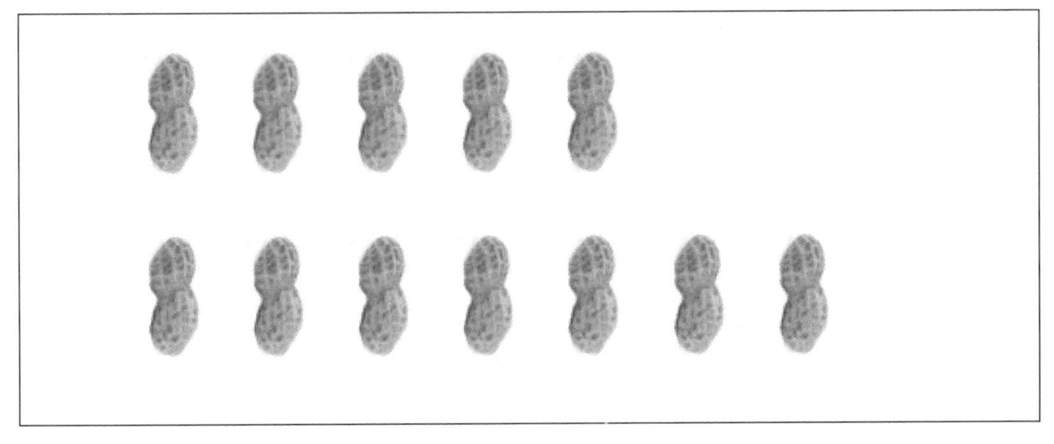

Fig. 2.11. Peanut task, part 1

Then as the children watch, point to the first row with your finger moving across the length of the row of peanuts. Then point to the second row with your finger moving across the length of the row. Ask the students, "Which of these two rows has more peanuts?"

One hundred percent of the students given this image selected the row of seven peanuts as the row that had more (see fig. 2.12). But looking deeper into what was observed on their papers, the students revealed other areas for further emphasis or extension. One child counted all the peanuts and got the incorrect answer of 13 but circled the second row. Another child did not count at all and said, "there are two more in the row" with another classmate saying, "two more sticks out."

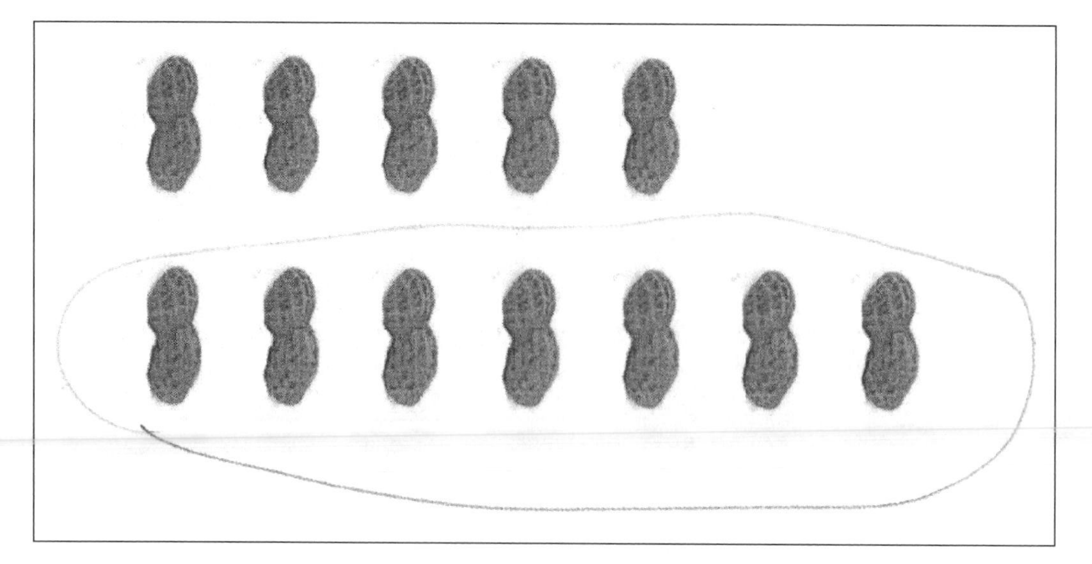

Fig. 2.12. One child's response to the Peanut task

Task: Rows of Peanuts, Part 2

Students are asked to compare an unequal number of peanuts in rows of equal length. They are asked to select the greater amount.

Immediately after the students answer part 1, show part 2 (see fig. 2.13) where students will compare an unequal number of peanuts in two rows of the same length.

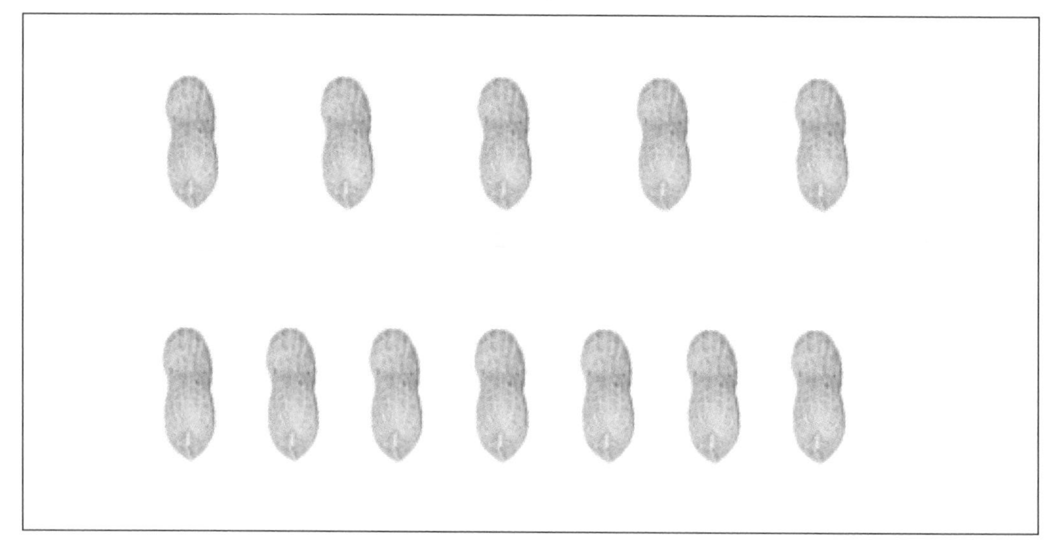

Fig. 2.13. Peanut task, part 2

Again, point to each row running your finger along its entire length. Ask students, "Which of these two rows has more peanuts?"

When the obvious visual discrepancy between the rows was taken away in the second problem, the students could not rely solely on the length of the row to determine which row had more peanuts. More than one-third of the students said the rows were the same as shown in figure 2.14.

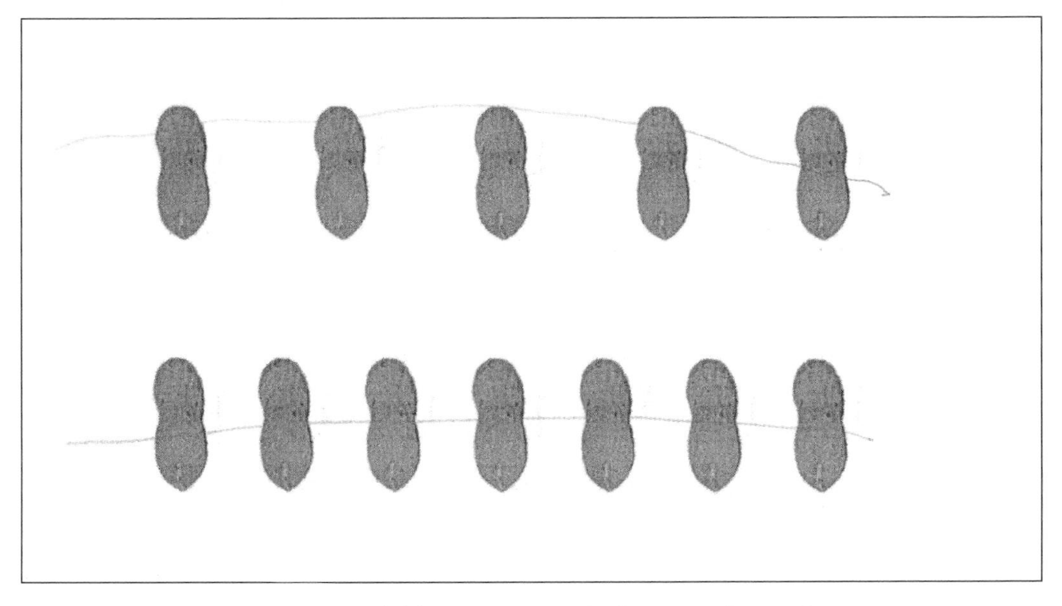

Fig. 2.14. A child's response to the Peanut task, part 2

This child stated, "I sketched it out. They are the same–19." Other children just said they were the same with no counting observed.

Whereas in the first problem children rarely counted, in this second situation, many children counted and some even offered their thinking, "It's hard to tell, so I had to count. Those were spread out and those were together. One has 7 the other 5." (See fig. 2.15. on the following page.)

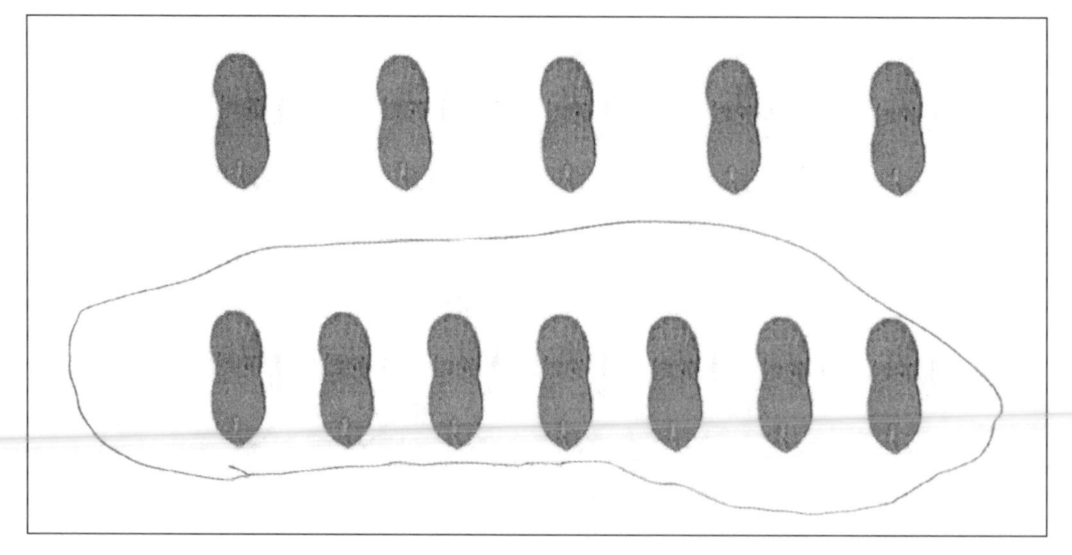

Fig. 2.15. A second response to the Peanut task, part 2

When the items in both rows are no longer vertically aligned (or equally spaced according to the same unit of distance), the children are challenged to use mathematical relationships of quantity. This situation demonstrates the point where students are counting as a necessary tool to compare quantities. You want to move children from merely making their decisions by length or by the space covered to making their decisions by comparing actual quantities of objects with an explicit focus on the attribute and unit used to determine an answer to "How many?"

Because children's skills accumulate as their number of experiences to think mathematically increases, these tasks can help assess children's early mathematical growth. This assessment should include teacher observations along with students' written work, and leads to better decision-making as to what interventions or next instructional steps should be taken either to fill in missing concepts or to challenge the child further on the basis of identified strengths. As you note from the children's dialogue we shared, just looking at students' papers does not reveal the whole story of what they know.

Development of the number line

The concept of *number line* begins with preschoolers as a discussion of a counting model called a number path (Fuson 2009; NRC 2009). Similar to the trail on a familiar game board, the children can begin to see a number path as a set of equal-size units that they count (Siegler and Ramani 2009). By setting down same-size 4 × 6 index cards numbered from one to ten (see the first image in fig. 2.16) and then having students walk on the cards as they count, the set of number names they experience is the initial step to their eventually interpreting the number line. This path can also be modeled with same-size colored Cuisenaire rods to build a number path on a table (Dougherty et al. 2010). Later, when students are familiar with the number path, you can give them a task where a blank card is positioned on the path; they then must determine the card's missing number. Start with one blank card, and then add one or more blank cards whose numbers need to be identified.

When students can move on the number path by counting on the amount from their current position, rather than going back to the start position, they're ready for the next developmental level, a walk-on number line on the floor already marked with lengths and numbers. Have children observe how the card units fit between the "steps" on this path, starting at 0. It is important to discuss the relationship between the number of units on the path and what the tick marks and associated numbers represent—the end of each unit and the count of the units up to that point, respectively. This action supports children's understanding of cardinality. Eventually the numbers on the number line will be a measure of how many units there are from the 0, used as the origin, or starting point, to that position on the line in a continuous model (with arrows on either end of the line). At this stage, discourage students from counting the tick marks because it undermines the concept of unit and obstructs the goal of having students see the number as marking the end of a step, space, or unit. Children will begin to recognize that the number is an indication of the measure of the number of units up to that point. This focus on distance from the 0 is why all number lines should have a 0 rather than starting with 1, using the 0 as a record of where children start in the number count of whole numbers. And, without a 0, it is not clear to young children what unit is being used to create the number line. This progression to marking the end of the unit when we count (making sure that the discussion is around the length of the unit, not the area [the space covered]) highlights the importance of the attribute because students can sometimes focus on the wrong attribute. Below is a set of number lines, beginning with those focusing on the actual tiling of units on a path to the more advanced use of empty number lines that will support development of more complex and abstract thinking to represent operations in grades 3–5 (see fig. 2.16).

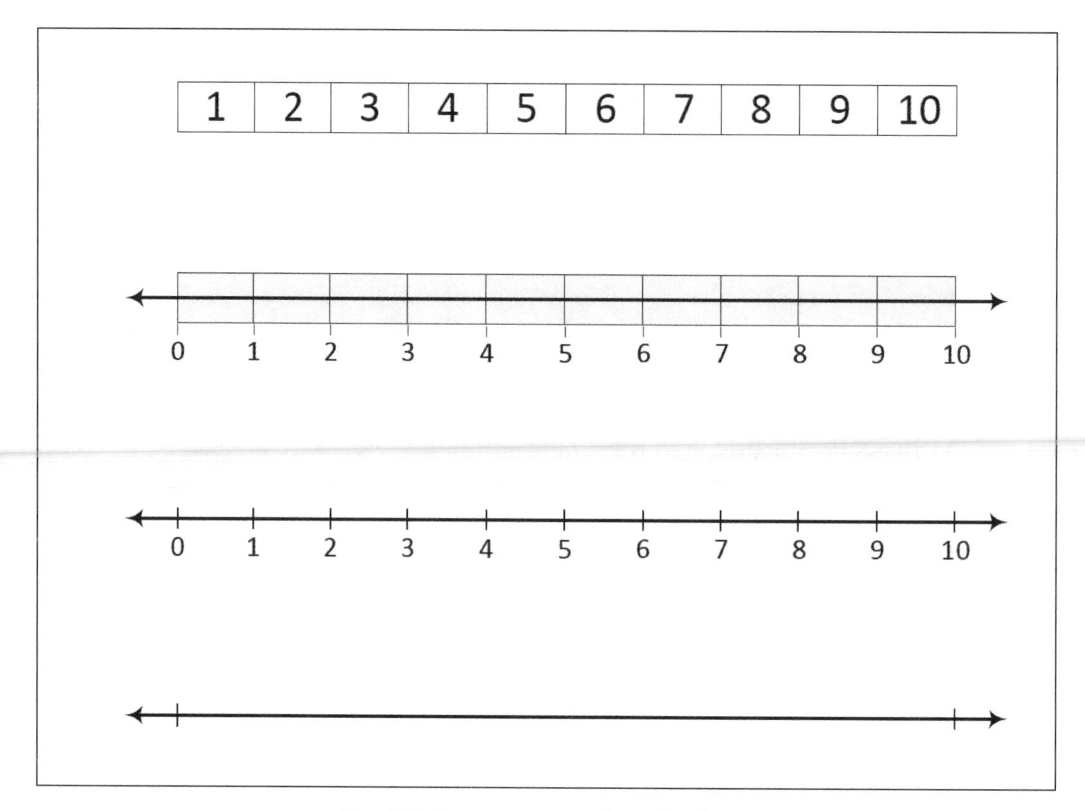

Fig. 2.16 A progression of number lines

Having students locate numbers on an empty number line is a worthy task and a useful predictor of future mathematics performance (Booth and Siegler 2008; Geary 2011; Sasanguie et al. 2012). But to demonstrate the understanding of the relative position of numbers, children must be aware of number—both the cardinal representation and the identification of the numerals (Gifford 2014). This awareness begins to develop when children discover that there is a relationship between numbers and space (Bueti and Walsh 2009). A useful task aligning with the investigation of number and spatial representations is estimating the position of numbers on a number line marked with only reference points or anchors. For example, start with a number line that only shows the positions of 0, 5, and 10—with each marked by the numeral and a tick mark. Ask students to show where other numbers belong (e.g., 3 or 9) and discuss their reasoning of why they put the number in that space. Later, students can move to such anchors as 0, 5, 10, 15, and 20 or to another option such as a line marked with 0 and all decade numbers ending at 100. The more proficient students are at counting forward and backward (particularly in steps of 2 and 3), the stronger their ability to accurately estimate the location of the number in these activities (Ebersbach, Luwel, and Verschaffel 2015). As students become more sophisticated in reasoning out the location of the numbers, the amount of concrete supports (such as walking in steps on the line) as well as the use of anchors should decrease, eventually moving to positioning numbers using only such anchors as 0 and 100.

As students progress from counting units verbally as they walk the "concrete" number line to drawing number lines on paper with numbered tick marks, the expectation is that children will eventually develop a mental number line (Dehaene et al. 2008). Siegler and Booth (2014) suggest this greater sophistication will commonly happen by age six. The development of a mental image of a number line can be prompted by having students estimate where a given number is positioned on a blank number line with anchors marked (as suggested above). Another option would be to write an X on a blank number line marked with beginning and end points and asking students which number (possibly from a set of three) is located in that space. Using estimation tasks with this more abstract number line representation will support children's broader mathematical skills and conceptual understanding (Schneider et al. 2018). This development of a mental image of a number line could enhance such topics as number sense and place-value concepts, and can eventually support such topics as rounding, the magnitude of fractions, and the understanding of positive and negative integers.

Summary: Learners, Curriculum, Instruction, and Assessment

To effectively teach the mathematical ideas presented in this chapter, teachers must have knowledge of the four components (learner, curriculum, instructional strategies, and assessment) presented in the Introduction. In the following sections we summarize some key ideas for each of these components.

Knowledge of learners

Before children enter formal schooling, they come with ideas about numbers and counting. This chapter explores how the idea of *unit* is critical if students are to understand number relationships. Their attention in the early grades to attributes that are not relevant to the determination of a count can hamper their ability to see relationships between the count and the unit that determines the count, as in the snake task described earlier. We questioned children about the two snakes' seemingly counterintuitive measures, not to confuse them but to cement their understanding of the importance of the size of a unit in thinking about which length is longer or shorter both visually and by a measurement. The understanding of what the unit is in each situation is a building block for solving problems and the mathematical ideas to come. As we see children gaining skills from physical situations and the use of number words to symbolic skills with numerals, we also see the need to carefully bridge these representations with an emphasis on the unit.

Knowledge of curriculum

The content of the curriculum and the order in which it is shared must prioritize particular essential understandings and ideas. Engel, Claessens, and Finch (2013) studied kindergarteners in the fall of the year and found that 95% of the students had mastered basic mathematical topics such as counting and the recognition of shapes, but teachers were found to spend approximately thirteen days a month repeating these same ideas. Instead of repetitive instruction, we suggest examining critical concepts that are new to students in depth. Though the idea of unit is mentioned in mathematical practice 2 in the Common Core State Standards, a formal consideration of the chosen unit in any given situation is not as consistently highlighted as a specific area of study as it should be in current curriculum materials for young children. Recognizing the early need for an explicit focus on the unit will pay off in important ways as young learners develop from basic counting to units of tens and hundreds and to older students using a number line marked with hours, dollars, miles, or other units to think about a variety of problems, including elapsed time, costs over time, distances on a trip, and so on.

Knowledge of instructional strategies

Aligned with the findings of Gersten et al. (2009), we have suggested explicit instruction (not direct instruction) in supporting young learners' ideas and conceptual understandings leading to number sense. The use of the combination of concrete and semiconcrete visuals and symbolic representations (CSA) as instructional strategies has been known to support learners; this chapter aligns with that approach. For example, number lines help students order numbers as they see the regularity in the pattern of the magnitude of numbers as larger numbers come later in the oral count and on the number line moving from the left to the right and vice versa for smaller numbers. These connections between concrete experiences and more abstract representations strengthen children's grasp of number concepts such as the unit.

Here are four initial steps for introducing the number line as a representational strategy in mathematics instruction for young learners. During math instruction, build in multiple opportunities to—

1. Encourage students to count the units on the number path out loud as they step to each location.
2. Link the use of the number path with a number line (with arrows on each end) of units the same size as the number path spaces; connect the end of any unit in a given set of units with a tick mark and its corresponding number. This shift is substantive because children must modify their thinking from counting numbers of objects (unit cards on the number path) to units of continuous length (a distance shown on the number line). Ask students to verbalize their thinking.

3. Ask questions that support students thinking about number relationships, and challenge them to use multiple representations, if needed, to compare magnitude, for example, "Which is larger, 9 or 5? How do you know? Can you show me on the number line?"
4. Consider basic addition or subtraction problems by having students combine or compare two lengths.

The consistent use of the number line helps students think through and reason about problems, not just as young students but even later in their mathematical learning when number lines can help combat misconceptions, such as the misapplication of whole number knowledge to fractions (Hamdan and Genderson 2017).

Knowledge of assessment

At this stage, students' assessments should rely on the powerful process of "kid-watching," or observation. This is where students' comments about their recognition of the unit, their walk on the number path, and their reasoning about the size of numbers are made visible. By asking students to identify the larger portions of cookies and other tasks in this chapter, the activity is, in fact, the assessment. Task selection is critical. Talking to students and noting their reasoning and sense making can help tailor instructional next steps to each student's strengths. Even identifying confusion, such as in the conservation tasks, gives insight as to children's thinking and their need for more experiences with specific types of problems.

Conclusion

Chapter 2 explored the essential idea of unit. This foundational mathematics concept is often hidden within the mathematics lesson; it needs to be made central and prominent in children's conversations about numbers and their relationships. We start with grounding students' understanding with quantitative comparisons without numbers, and then move to the Number core with a look not only at the numerals and their values but also their connections to quantities. Finally, we bring the discussion to number relationships in the Relations core.

As you can see in this chapter, we need to look beyond these early learning years and their specific standards to be met. Instead, we need to adopt a long-term view of how ideas such as unit or strategies such as using the number line to support thinking play out over time. By emphasizing long-term approaches relevant to the present as well as to the mathematics ahead, we can coherently plan the overall approach to a child's mathematics progress.

Chapter 3 Counting

Big Idea 3
Meaningful counting integrates different aspects of number and sets, such as sequence, order, one-to one correspondence, ordinality, and cardinality.

Essential Understanding 3*a*
The number-word sequence, combined with the order inherent in the natural numbers, can be used as a foundation for counting.

Essential Understanding *3b*
Counting includes one-to-one correspondence, regardless of the kind of objects in the set and the order in which they are counted.

Essential Understanding *3c*
Counting includes cardinality and ordinality of sets of objects.

Essential Understanding *3d*
Counting strategies are based on order and hierarchical inclusion of numbers.

Getting to Counting with Understanding

When we think of counting, we often think of children reciting numbers in order, 1, 2, 3, 4, and so on. *Rote counting* is reciting numbers in order from memory without any reference to a set of objects. There are, however, other types of counting that are important in number development.

One-to-one correspondence, or finding the *cardinality* of a set of objects, is a different type of counting that requires students to assign a unique number name to each object counted. These number names must be assigned in order (1, 2, 3, 4, and so on), but the order in which the objects are to be counted does not matter. The last number name given to the last object counted is the cardinality of the set of objects, that is, the last number named is the total number of objects in the set.

Ordinal counting uses a different set of number names and assigns an order to the objects being counted. Children must still count in a one-to-one fashion, but the number names they give to an object are indicative of the order of that object in a set. These number names are first, second, third, and so on.

Each type of counting is important in the development of a child's understanding of number. The differences between the two methods of counting are the focus of Reflect 3.1.

Reflect 3.1

Which type of counting is the most difficult for children you work with? Why?

Number-word sequence can be used as a foundation for counting

Before students can successfully carry out any computational problems, the ability to count reliably is key. One way to explore a child's counting ability is to carry out counting tasks with a character who sometimes counts properly and at other times makes mistakes. The capacity to point out counting errors is an important step in helping children refine their own counting skills.

Task: "Mixie Up" Messes Up

Students point out the counting errors of Mixie Up, a puppet who sometimes counts accurately, and sometimes not.

"Mixie Up" is a puppet that sometimes counts correctly and at other times just never quite carries out the counting process accurately. You will need to give children advance notice that Mixie Up sometimes gets confused and they must watch her carefully and point out when she is counting correctly. When she is making mistakes, they need to name the mistake and suggest how to help her.

This task requires a puppet or stuffed animal and a set of cubes or objects that can be easily seen (nothing flat). Introduce the puppet as a friend who is just learning to count and needs some assistance from the children. Tell the children that Mixie Up might make some mistakes because she is just getting started in counting and they need to gently tell her when she does something they want her to correct. In Reflect 3.2, as you consider how correcting counting mistakes can be productive for children learning to count, you might also think of what types of mistakes would be most constructive for them to correct.

Reflect 3.2

Why do Mixie Up's different mistakes support your students' mathematical understandings?

Have the puppet count in a variety of incorrect ways. Some examples are listed below:

- Skipping objects while still stating numbers or stopping the number count early and not counting all the objects at the end of the row (lack of one-to-one correspondence)
- Counting an object twice, that is, double counting (lack of one-to-one correspondence)
- Counting out of the conventional sequence, for example, 1, 2, 6, 4, 8, 5 (lack of understanding of an accepted convention in the sequence of numbers)
- Counting and not being able to answer the question "How many do you have, Mixie Up?" (lack of understanding of the cardinality principle)
- Asking a child to show a group of a specific number of cubes (explores child's ability to use counting as a tool)
- Comparing two groups of objects by counting them (comparing sets)

Children who identify the problems with Mixie Up's counts love to tell exactly what went wrong and then state the correct count. Here are some children's thoughts about Mixie:

- "She didn't say the exact right number every time."
- "She did it wrong as she wasn't saying the number on each cube" [*and then the child proceeded to model the right way of counting the cubes by touching each*].
- "She said she had four at the end but she has six. See."
- "She needs to learn how to count."

- “She went backwards. Instead of saying seven, she said five! I want her to always count forwards all the way to this one [cube].”

This role of student as teacher helps many children cement ideas that they are forming about the process of counting.

As you may have noted, the puppet is asked to count either to find the total number of objects or to find out whether every object in one grouping has a corresponding object in another group. You may notice that young preschoolers tend to count when asked how many are in a set but often do not rely on counting when asked to tell if two sets are the same amount using one-to-one correspondence or to create a set with a given number of objects.

You can extend the original puppet counting tasks for children by exploring the work of Sophian (1988). For example, the puppet can give a total number of objects in two sets of objects (extension of the *how-many* task) or you can ask the puppet to judge comparisons (greater than or less than) by exploring the relationship between two subsets (the *compare-sets* task). Difficulty can be adjusted several ways: the puppet can count all of the objects together (*counting all*) or counts the two groups separately (*counting subsets*); the problem involves small groups (5–6 total objects, with 2–3 objects per subset) or larger groupings (11–12 total objects, with 4–8 objects per subset); or the two groups are equal or unequal, with one group having one or two more than the other. Having the mixed-up puppet identify two groups as equal when they are not or say a group is greater than another set when it is not, for example, is an important way for children to begin to understand these concepts.

Because the use of physical connection reduces errors (Bashash, Outhred, and Bochner 2003), children should be encouraged to touch and move the items they have counted as a way to organize the process. In the act of moving the item after counting it, students more easily keep track of their count. Any child having difficulties with accurate counting should be an encouraged to practice this one-hand method (van Klinken and Juleff 2015).

Other diagnostic tasks for individual students who may not have shared ideas in the larger group might include counting forward from one, counting a small set of cubes or counters, or counting out requests for specific numbers of counters. These diagnostic tasks can pinpoint children’s strengths, which will allow for a more tailored and targeted intervention, whether it’s a reexamination of foundational ideas or an extension to more advanced counting skills.

All students, of course, will benefit from mastering more advanced counting skills such as–

- counting on orally from any number under 100 (the teacher gives the starting number for the count);
- saying the number before or after any number (as on a hundreds chart);
- counting on orally from a numeral such as 101, with the teacher presenting the starting number for the count either orally or by showing the written numeral; and
- counting backward from any number (the teacher presents the number either with a written presentation of the numeral or a verbal reading of the number).

Tasks such as these advanced counting opportunities help determine whether children have a more sophisticated knowledge of the system of natural numbers (Rips, Bloomfield, and Asmuth 2008).

Finger counting

Children naturally touch objects as they count, which reinforces their counting skills. Researchers have found that the increased knowledge and proficiency gained by using their fingers ("finger sense" [Gerstman 1940]) to count improves children's performance in mathematics (Berteletti and Booth 2015; Fischer 2008; Gracia-Bafalluy and Noël 2008). Some researchers have found a brain connection, or overlap, between the finger touching and children's number representation (Penner-Wilger and Anderson 2013).

Fingers often come into play when children learn to count, often starting naturally by using the index finger to count one object or suggest they are one year old. Over time, the order of finger counting shifts to starting with the thumb or little finger. Then, the fingers are ticked off one at a time as the counting sequence continues, and the child responds with lifting fingers in a one-to-one correspondence in a logical progression across the hand. Finger representations of counts are powerful models of one-to-one correspondence that support the child's thinking while linking two components–physical elements and symbolic by means of verbal systems. The physical number of human fingers aligns to the base-ten system, such that when ten fingers are counted, there needs to be a new grouping of the fingers to start anew. There are connections that go beyond the common use of finger counting, such as the English word *digit,* which represents one finger as well as a single number. Fingers help young children keep track of discrete objects, buttressing one-to-one correspondence and equivalence–ideas needed for future mathematical concepts.

Counting includes ordinality of number

Certainly, one way to not only motivate young students but also engage them in problem solving is the use of context. Using common real-world situations or contexts based on stories from children's literature can provide a source of interest and even stimulate the child's storytelling, which can lead to greater understanding of the structure of word problems. Let's look at a scenario of a train stopping at multiple stations.

Task: People on the Train

Children deal with ordinal numbers while also counting "people" getting on and off a train.

Give students a collection of craft sticks with simple faces drawn at the top of each and a series of train cars (see fig. 3.1). Tell short stories of people getting on the train in the first, second, or third cars. You can also designate how many are getting on or off each of the cars as the train makes a series of stops. Then, ask the students to tell a story that another child will demonstrate, putting people on the cars and following directions to place four people in the second car, one person in the fifth car, and so on. Students can count the total number of people on the train at the various stops after people get on and off. For example, one child said, "I had three in that first car and three in the other car and one more. Three and three is six and one more is seven." The People Train is a meaningful way to practice counting and to think about ordinal numbers. It also connects stories to mathematical ideas. Delve deeper into the complexities of the People Train task in Reflect 3.3.

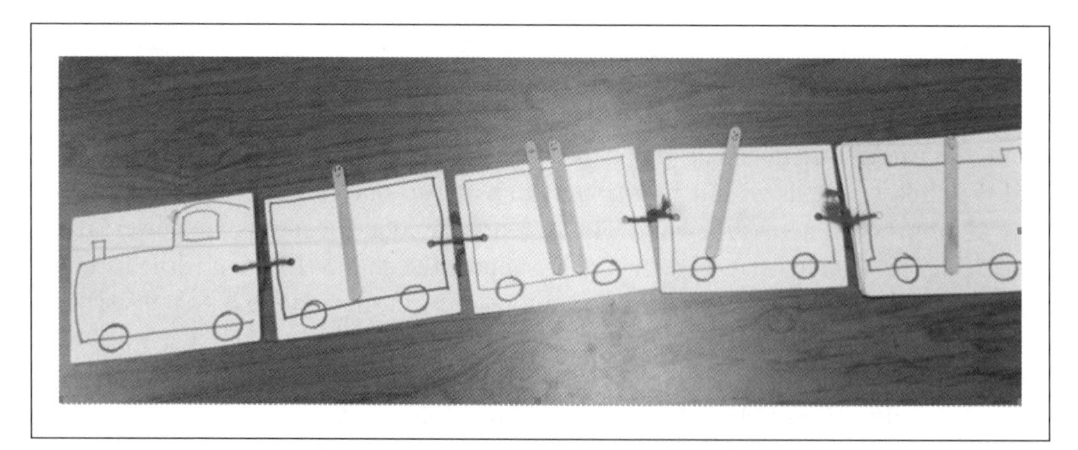

Fig. 3.1. People and train materials

Reflect 3.3

How does the People Train task help students develop their understanding of word problems? How does their desire to either get more people on the train or off the train support their thinking about whole number operations?

From Subitizing to Counting

Crosby and Sophian define *subitizing* as "virtually errorless apprehension of very small numbers of items" (2003, p. 1600). When children subitize, they must visually absorb multiple objects in an image immediately and simultaneously. The mental tension between trying to count by ones and visually scanning the display of objects is a demanding task for a child. But, this skill is another of the "main abilities young children should develop" in the collection of early mathematical learning that predicts children's later understanding of mathematical ideas (Clements and Sarama 2014, p. 10).

Images such as dot cards or five- and ten-frames help children develop this skill; given a pattern of five dots in the positions of those on a standard die, some children just "see" five. Other children may see two dots on the left, two dots on the right, and one dot in the middle. These approaches that discourage counting-by-ones clearly support the beginning of children's ability to compose and decompose numbers.

There are many tasks that not only develop children's skills in subitizing but also assess their progress as they gain skill in grasping the orientation of dots or counters and linking those amounts to a quantity. When the child must pull back to count individual dots, it could be an indication that they are only able to see a number as part of a series.

Task: Dot Plates

Students practice "seeing" numbers as patterns composed of individual items.

Gather inexpensive paper plates and two colors of adhesive dots (available in stores with office supplies). Start with simple dot patterns; then move to patterned sets in which a doubled amount or a decomposed part-part arrangement can be identified (see fig. 3.2 for possible options). Hold a plate up for four seconds and ask children to focus on patterns (not counting by ones) to identify how many dots are on the plates. To highlight the amount as the combining of two smaller parts, you can use two colors for the dots. (Two colors of dots on a plate can also support ideas about addition, as noted in Reflect 3.4.) Students can name the number verbally, and write the corresponding numeral down on a wipe-off slate or use counters to model the amount on a paper plate at their desk. For students who may need additional adaptations, you can give them a collection of identical dot plates as a partner so they can find the matching plates.

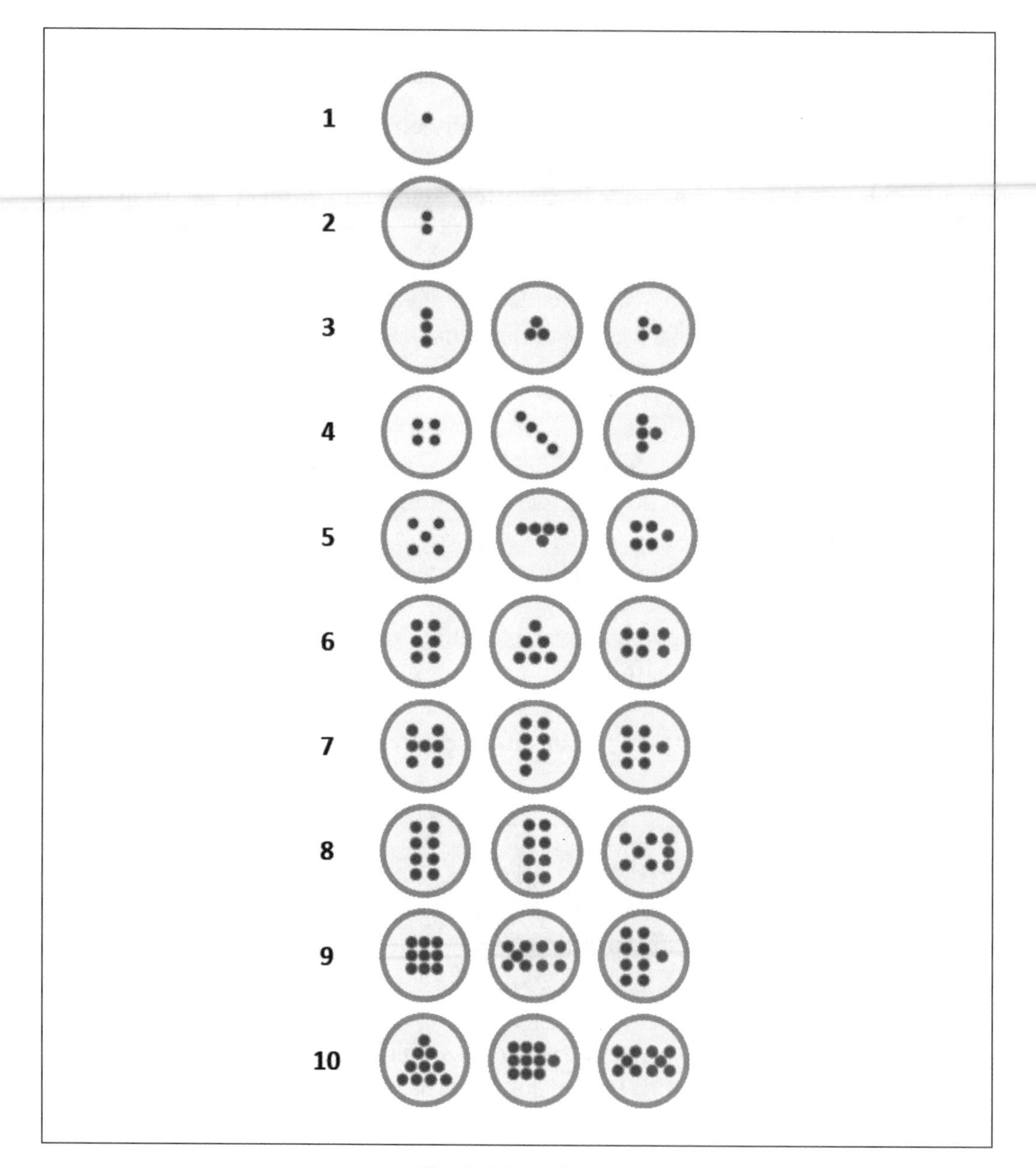

Fig. 3.2. Dot plates

Reflect 3.4

How can the plates with sets of two different colored dots support students' thinking about addition?

Task: Five-Frame or Ten-Frame Blink

A quantity of dots on a five-frame is flashed for three seconds for students to determine the number of dots.

Students will not want to blink, or they will miss the amount of counters in this activity. Starting with a five-frame (see fig. 3.3), flash a quantity of counters on a five-frame using the document camera for three seconds. Ask students to explain how they saw the counters. They might notice patterns or they may observe what is empty in the row. (Note that there is a convention with five-frames and ten-frames of placing the counters left to right and filling the top row before going to the second row.) When students graduate to ten-frames (see fig. 3.4), they can also think about what is needed to make a ten. Ask, "Here's seven. How many more will add up to ten?" Double ten-frames are a good extension, too!

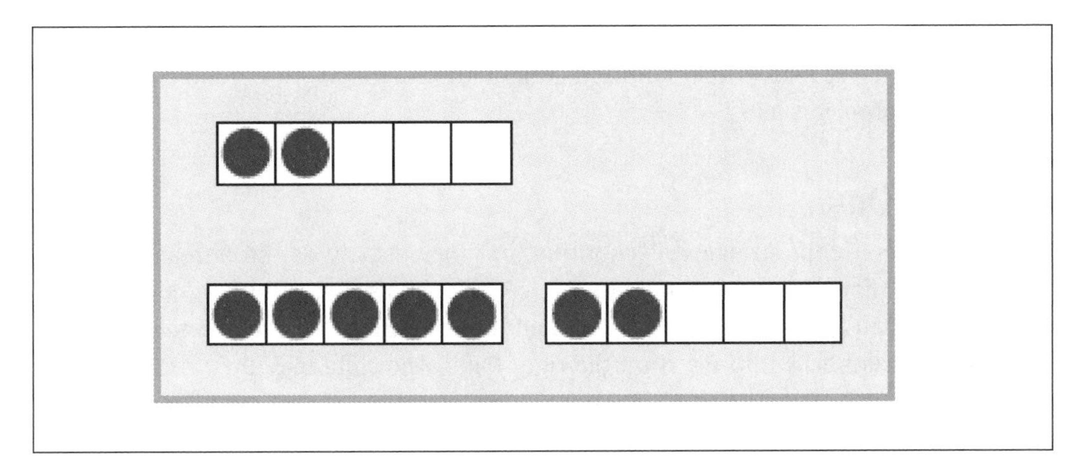

Fig. 3.3. Five-frames with possible amounts shown for the Blink task

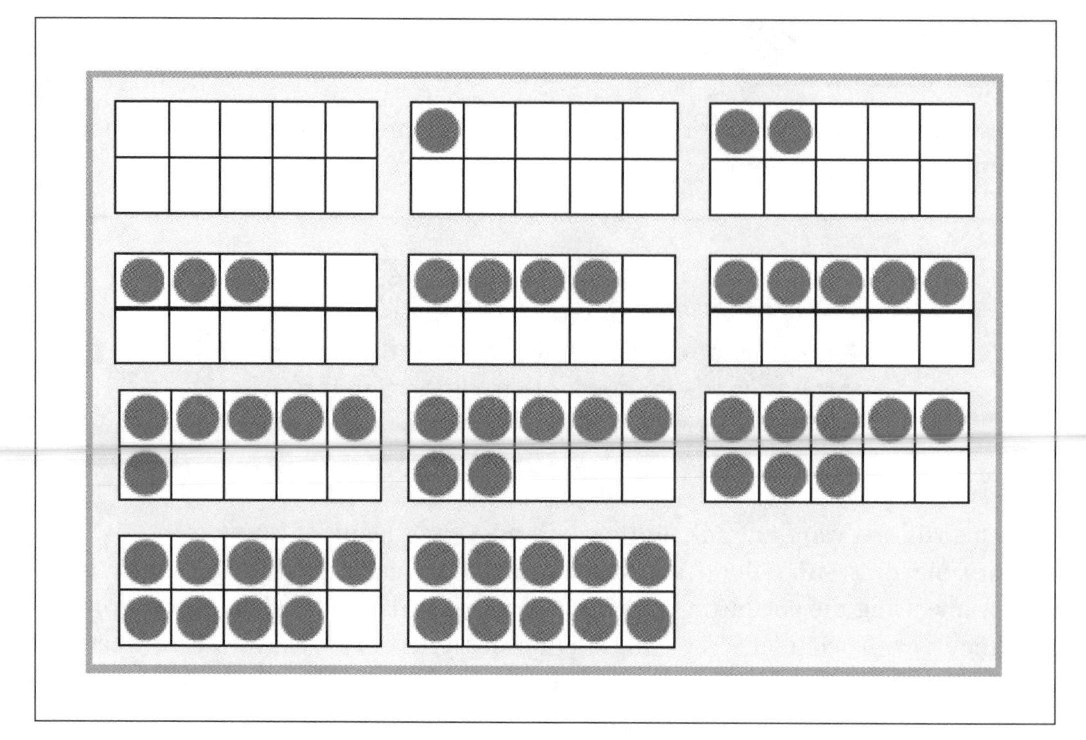

Fig. 3.4. Ten-frame collection for the ten-frame Blink task

The ten-frames can also be used to assess basic addition skills: flash two of the frames in figure 3.4 and ask students to write the total. Here they will need to capitalize on their ability to use the combinations that equal 10. From those facts, students can derive many other addition combinations, which will later lead to solving such problems as $17 + 3 = \_\_\_$.

Counting On

Before students use the strategy of "counting on," they usually go through a stage of "counting all." For instance, when adding 2 + 3, the child's first count might be "one, two," raising two fingers on the left hand, and then counting, "one, two, three," matching the verbalizations with fingers on the right hand. The child may then recount the extended fingers by even touching the nose as the count progresses with "one, two, three, four, five." This is a *count all* approach. As you observe your students, notice whether this strategy is being used, particularly if the children are older than five or six years. If so, the children need to move away from that approach to *counting on* to join quantities or to *counting back* to separate or take away amounts.

Counting on is a critical component to strategizing addition, particularly mental addition. Start by using the puppet Mixie Up to count on from six, adding three more cubes. Set all the cubes on the document camera and have Mixie Up hover over the six cubes. Using a prolonged sounding out of the word six as in "siiiiiiiix," have Mixie Up then say "7, 8, 9" as the puppet continues the count from six and touches the other cubes. This approach uses a "think aloud" technique to make Mixie Up's thinking visible to the children. Then ask the children to try to count on from other numbers using the same approach.

Counting back can be very challenging. You can experience this level of challenge yourself if you try to count back in a foreign language that you are not fully fluent in. Keeping the count sequenced properly while trying to remember the number words is a true test of working memory. Many children find it demanding to count back any amounts greater than three (Sarama and Clements 2009).

Summary: Learners, Curriculum, Instruction, and Assessment

To effectively teach the mathematical ideas presented in this chapter, teachers must have knowledge of the four components (learner, curriculum, instructional strategies, and assessment) presented in the Introduction. In the following sections we summarize some key ideas for each of these components.

Knowledge of learners

Learning to count collections of objects to determine cardinality presents some interesting challenges for young children. They must coordinate the use of a unit and use that unit to count a set that may not be arranged in any logical fashion. Either mentally or physically, children have to reorganize the set so that the counting is accurate. Additionally, they must recognize that the final number they say represents the count of the set. The coordination of these components is not trivial.

Ordinal counting adds another complexity to the task. Whereas the order of the objects when counting for cardinality does not matter, the order becomes highly significant with ordinal counting. The arrangement of the objects has to be coordinated with the count of the objects. Thus, some of the strategies students used for cardinality may not be appropriate for ordinal counting.

The distinction between the two types of counting is yet another layer of sophistication. Not only are the counting techniques different but young children must also determine when to use each type. They must first interpret the context and then reply appropriately.

Knowledge of curriculum

There is substantive information that not enough of the preschool curriculum focuses on mathematics instruction for early learners (NRC 2009; Piasta, Pelatti, and Miller 2014). Just sharing "calendar counting" activities is not enough for young learners to reach the critical standards, which include kindergarteners counting to 100 by the end of the school year. Activities that purposefully support students' counting, including one-to-one correspondence and subitizing, are essential. But a scripted curriculum that marches at a rigid pace or direct instruction aren't the desired approaches to early childhood mathematics instruction. Time for intentional explorations and interactions with number relationships by playing mathematics games with partners or through activities with materials, both individual and collaborative, are more developmentally sound approaches (Clements and Sarama 2014). Each day, starting in preschool settings, time must be set aside for mathematics instruction. This scheduling does not exclude the integration of mathematics instruction throughout the day.

As shown in research, the curriculum should include verbal counting, one-to-one correspondence, cardinality, subitizing, set comparison, and story problems as connected to the numeral identification in each component (Purpura, Baroody, and Lonigan 2013). This progression points to the centrality of the verbal counting component and, not surprisingly, to the research that identifies children's performance on that skill as a way to forecast future mathematics performance.

Knowledge of instructional strategies

Working closely with parents and families to share strategies you are using is an important first step to integrating mathematics into children's "real lives." Meet with parents and guide their mathematical interactions with their children by sharing how they can expose their children to a home environment where mathematics is pointed out to them in all sorts of real-life situations. Active engagement in the home with mathematics, a subject mainly taught in school and perhaps not discussed at home as often as other subjects, will enhance the mathematics in the formal school setting in ways that cannot be measured. Coaching families to use at-home teaching strategies can help overcome any differences in children's background in early mathematical ideas, which can result in higher levels of kindergarten readiness. Explaining the intentional goals of the activities as well as the mathematics behind the strategies supports the sense that we are all on the same team from birth to beyond schooling years. Responsible partnerships established between teachers and families in the early years can play out into long-term arrangements in which the work is shared and both are supported in the process of helping children shift from informal mathematics knowledge learned outside of the classroom to more formal mathematics which includes concepts, skills, and procedures taught in school.

Knowledge of assessment

The tasks presented throughout this chapter provide opportunities for assessing your students as they share their counting ability and their thinking about symbols and quantities. These tasks can be part of a set of formative assessments that collectively will provide evidence that supports your instructional decision-making and next steps for each and every child. Some of these tasks will unearth common misconceptions or challenges that can provide the data you need to take students back to more concrete approaches. Remember, it is important to identify students' strengths every step of the way so that you are not focusing on their gaps and weaknesses but building off of prior knowledge and understandings that can include their use of representations and their comprehension of concepts.

Additionally, simply screening children's counting skills by asking them to count as far as they can is a very informative assessment of their emerging numeracy. You can add depth to your understanding of students' counting by asking them to count from other numbers, such as requesting that they start counting from 17 or 32, depending on the level of challenge you want to provide. Using Mixie Up and seeing how well they identify her errors is also a good screener for several of the components of effective counting ability. Students should then be assessed for the ability to compare numbers, order a collection of numerals, and respond to acting out basic story problems with materials to represent the quantities.

Conclusion

From both the perspective of research and in every practical way, early numeracy, including counting, are critical life skills—abilities that for young children have been found to be a powerful predictor of their later computational fluency (Zhang et al. 2014), accuracy (Koponen et al. 2007), reading (Koponen et al. 2016), science and engineering (Claessens and Engle 2013), and general mathematics achievement (Purpura, Baroody, and Lonigan 2013).

into practice

Chapter 4
Building Ideas about Place Value

Big Idea 4
Numbers are abstract concepts.

> **Essential Understanding 4*a***
> Patterns in the number-word sequence provide a foundation for the abstract number concept.

Big Idea 5
A base-ten positional number system is an efficient way to represent numbers in writing.

> **Essential Understanding 5*c***
> The value of a digit in a written numeral depends on its place, or position, in a number.

Getting to Understanding the Base-Ten Number System and Place Value

Patterns in the number

The structure of number names plays a significant role in young children's learning of mathematics. Although you are accustomed to the way we write and say numbers in the Hindu-Arabic system, other countries use different systems for naming numbers that may make it easier for children to remember, as well as to support, their understanding of place value. Our structure gives English-speaking children a great deal of trouble starting with "teens" and continuing with some decade words. This relationship between language and mathematical ideas is complex and researchers are still exploring the connections (Moschkovich 2017).

This potential obstacle starts when young children encounter teen numbers in the range between 11 and 19 where they also face particular problems with the irregularity of 11, 12, and 13 (Ofsted 2013). There are complexities as well in Spanish for the numbers between 11 and 15. The U.S. system switches the place-value structure with these numbers as the tens place is spoken last instead of first, which reverses the pattern from most of the other numbers in the system. So, rather than saying the tens first as we would with 27–"twenty-seven"–which matches the numeral order, a number such as 17 is named the opposite way with the ones first–"seventeen"–an order that does not match the position of the digits in the numeral. However, many countries verbalize the numbers between 10 and 19 as ten-one, ten-two, ten-three, and so on. This more recognizable pattern in the language structure of their numbers initially allows for better understanding (Comrie 2005; Dowker, Bala, and Lloyd 2008). The result of our irregular system of number names in English requires significantly greater working memory, which is already fully employed when children are counting objects (Passolunghi, Vercelloni, and Schadee 2007). For children who are already trying to remember the sequence of the numbers or the count they are currently on or both, this disruption in pattern can be overwhelming. Not surprisingly, there are common mistakes, such as students saying "fiveteen" or writing fourteen as 41 or 410, after hearing the "four" in fourteen first (Thompson 2008).

Task: Count to Twenty

Young children practice the complexities of the number-naming system of the English language.

The basic task of orally counting to 20 can be given to young children. Here is an example of a first grader's count: "1, 2, 3, . . . 9, 10, 10 and 1, 10 and 2, 10 and 3, . . ." When asked how he was thinking about the counting, he referred to the stacks of connecting cubes he was using to count these numbers in class. The boy clearly understood his visual model of place value, but verbalizing the English numeral name was another matter. Reflect 4.1 focuses on similar difficulties with other number names.

Reflect 4.1

What other number names confuse children? Why?

After the number 20, the English system is more predictable, except for the naming of decade numbers. In English, our words for the decade numbers such as 20, 30, and so forth, do not precisely and easily match the pattern established with the ones. Whereas 60 is a straightforward match to "six-ty," 20 is not aligned with "two-ty." As with the teen numbers, some other languages have a more predictable pattern and logical sequence that links the decade number directly back to the ones digit in counting. For example, when counting in Chinese, the number 24 would be formed as the word "two ten four." This pattern makes counting considerably more regular for young learners in China (as well as in Korea and Japan). Instead of English words such as thirty and fifty, these language structures more effectively support not only place value but also the critical skill of decomposing numbers. For example, in the Japanese language, 287 is read as "two-hundred-eight-ten-seven," which aligns well with the break-apart strategy used in solving computation problems. Several researchers suggest this transparent linguistic pattern is one of the factors producing high performances in these countries (Ng and Rao 2010), but there are many other variables that play into these differences (Moschkovich 2017). However, there is evidence that kindergarteners in China respond three times faster than their U.S. counterparts on basic addition problems (Geary et al. 1993), which is a foundational difference that may be challenging to recoup. Preschoolers may, in fact, sometimes confuse the structure by saying "twenty-eight, twenty-nine, and twenty ten," not yet making the link to the base-ten system.

Rather than just teaching students to count using the format aligned with a place-value structure as some might suggest, children need to practice the complexities of our system. Here is a useful task that helps students understand the counting bridge between the decades.

Task: Crossing Over Game

This game helps young children learn to count from one decade to another.

This game helps students bridge the challenge of moving from one decade to the next in counting, and requires the game board and number cards found in Appendix 3. To begin, place a game board on a table that has __9 on the left half and ___0 on the right half (see fig. 4.1). Put the cards that have a 9 in the ones place in one pile and the cards with a 0 in the ones place in another. Children play this game in pairs; each child gets one of the card stacks (either the numbers that have a 9 in the ones place or a 0 in the ones place). One child chooses a number from the set of cards that ends in 9. That child (if playing alone)—or the partner—must find the card that follows in the count; for example, if 39 is placed on the game board, the child needs to find the 40 card to place on the other side of the board. You can also play by selecting a decade number and then finding the number that comes before it in the count.

Fig. 4.1. Game board and cards for the Crossing Over game

The Base-Ten Positional Number System

The base-ten system forms a new place-value unit by grouping ten of the previous place-value units.

Children who have challenges with understanding the basic concept of place value beyond the first grade will likely have serious difficulties with mathematical computations (Chan and Ho 2010; Hanich et al. 2001; Jordan and Hanich 2000). Grasping that the placement and positioning of digits in a number is the structure of our system is necessary for any advanced mathematical problem solving.

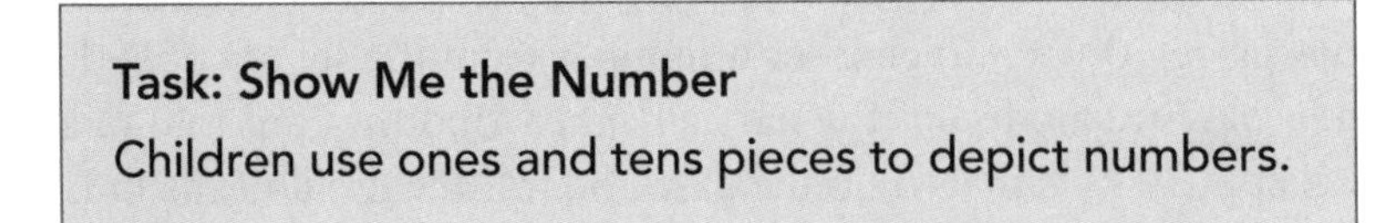

Task: Show Me the Number

Children use ones and tens pieces to depict numbers.

Have students represent numbers with tens and ones pieces from the collection of base-ten materials; for example, ask the students to create the number 22.

There are three levels of performance for this task (see fig. 4.2):

1. The child creates a unitary, or count-by-ones collection, by taking out 22 ones pieces.
2. The child creates the classic base-ten, or groups of ten, representation by taking out 2 tens pieces and 2 ones pieces.
3. The child creates an alternative equivalent, or non-standard grouping, where they take out 1 ten piece and 12 ones pieces.

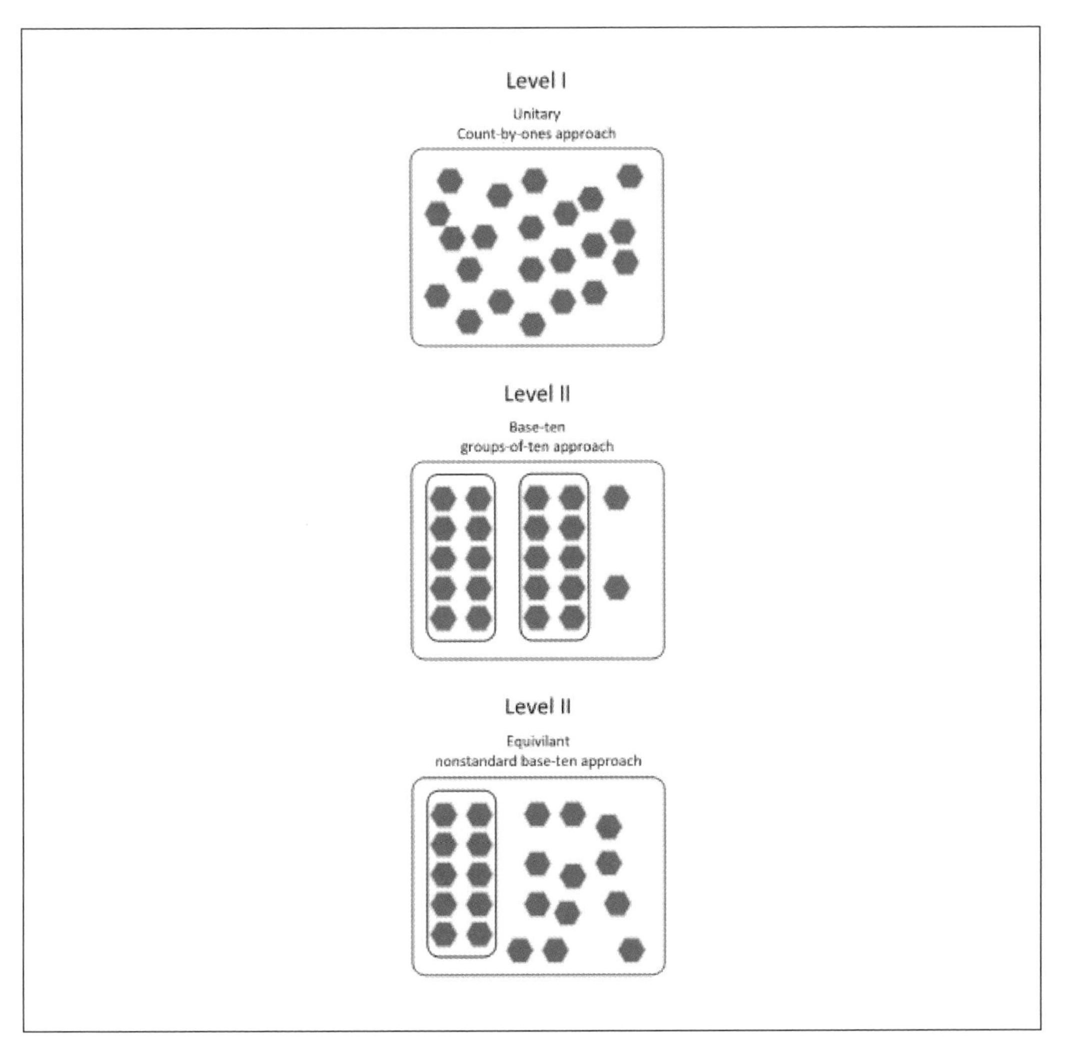

Fig. 4.2. Three levels of place-value understanding

Task: Birds on a Branch

Students count and organize "birds" in groups of ten.

In this task students are given an activity sheet with directions to figure out two things:

1. How many branches do you need to hold this group of birds?
2. How many birds do you have?

The constraint given is that only 10 birds are allowed on one branch, as shown in figure 4.3. If there are more than 10 birds, another branch is needed for the "extra" birds or each group of ten more.

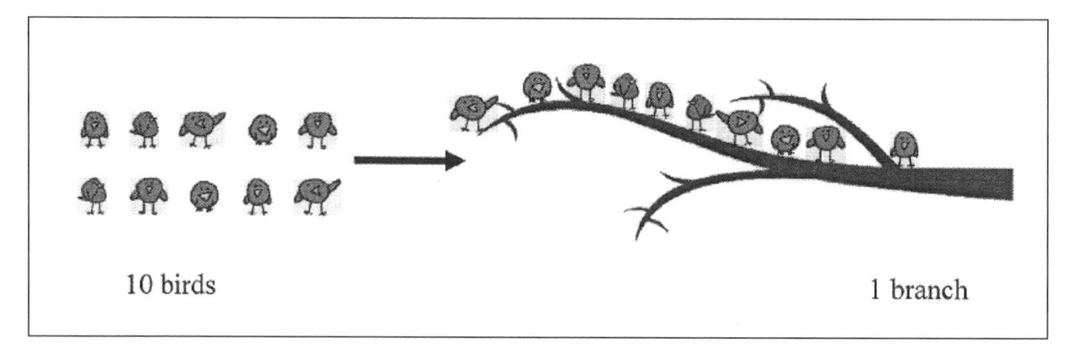

Fig. 4.3. Birds on a branch information

In the first problem, there are 12 birds. Here is a collection of student work on that case from a kindergarten class:

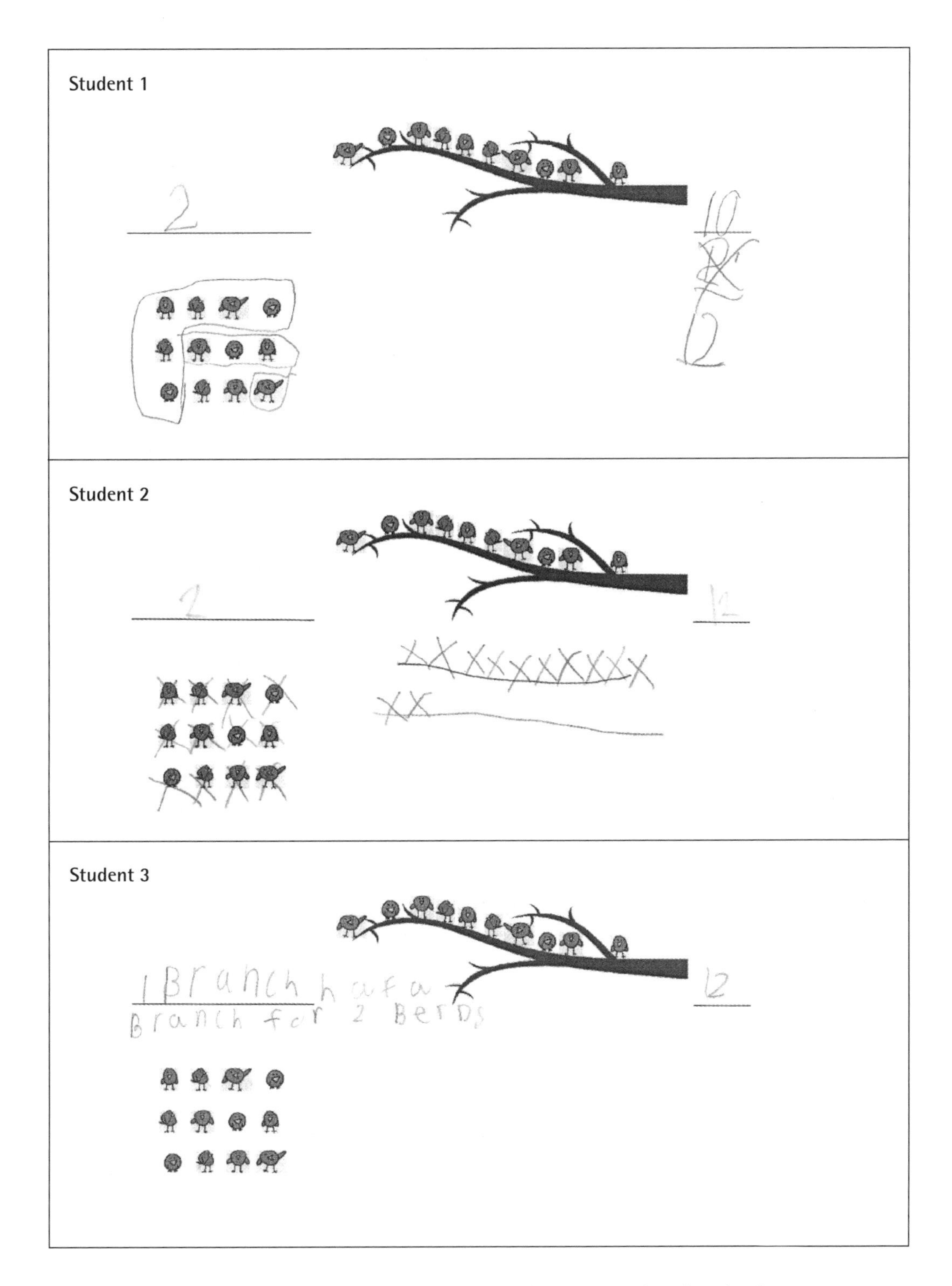

Fig. 4.4. Kindergarteners' work on Birds on a Branch with 12 birds

In this collection of student work samples from five-year-olds, you can see cases in which students circled the birds into a group of ten and then two more to add both parts together; a student who modeled the two branches by sketches that used a line and Xs to represent the birds; and a student who suggests a half of a branch is all that is needed for the two extra birds.

In the second problem, these students were given 16 birds. The higher number of birds resulted in more complications in students' thinking about the total amount and the arrangement on branches as reflected in the examples of student work below and on pages 67 and 68 (see fig. 4.5).

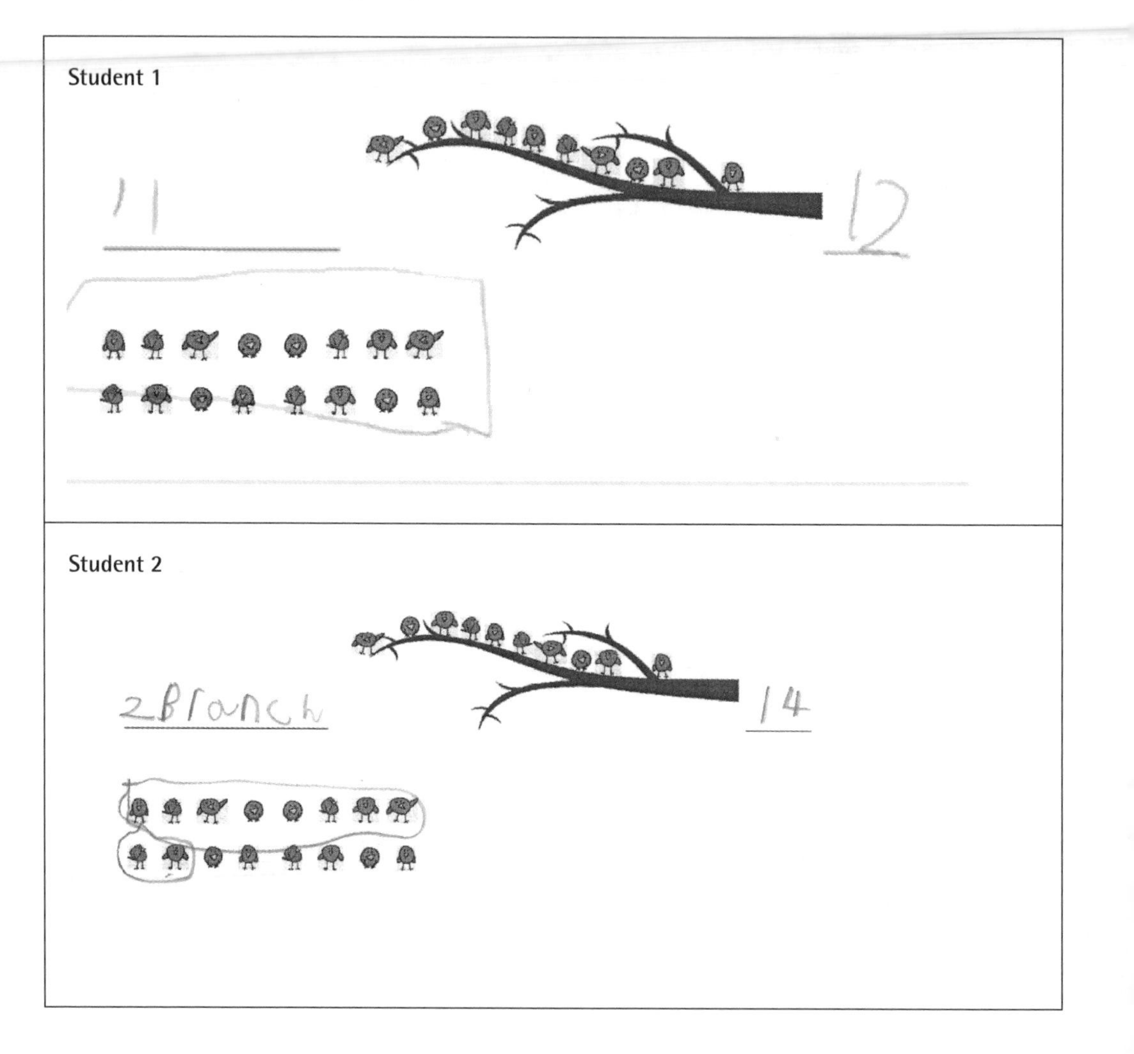

Fig. 4.5. Kindergarteners' work on Birds on a Branch with 16 birds

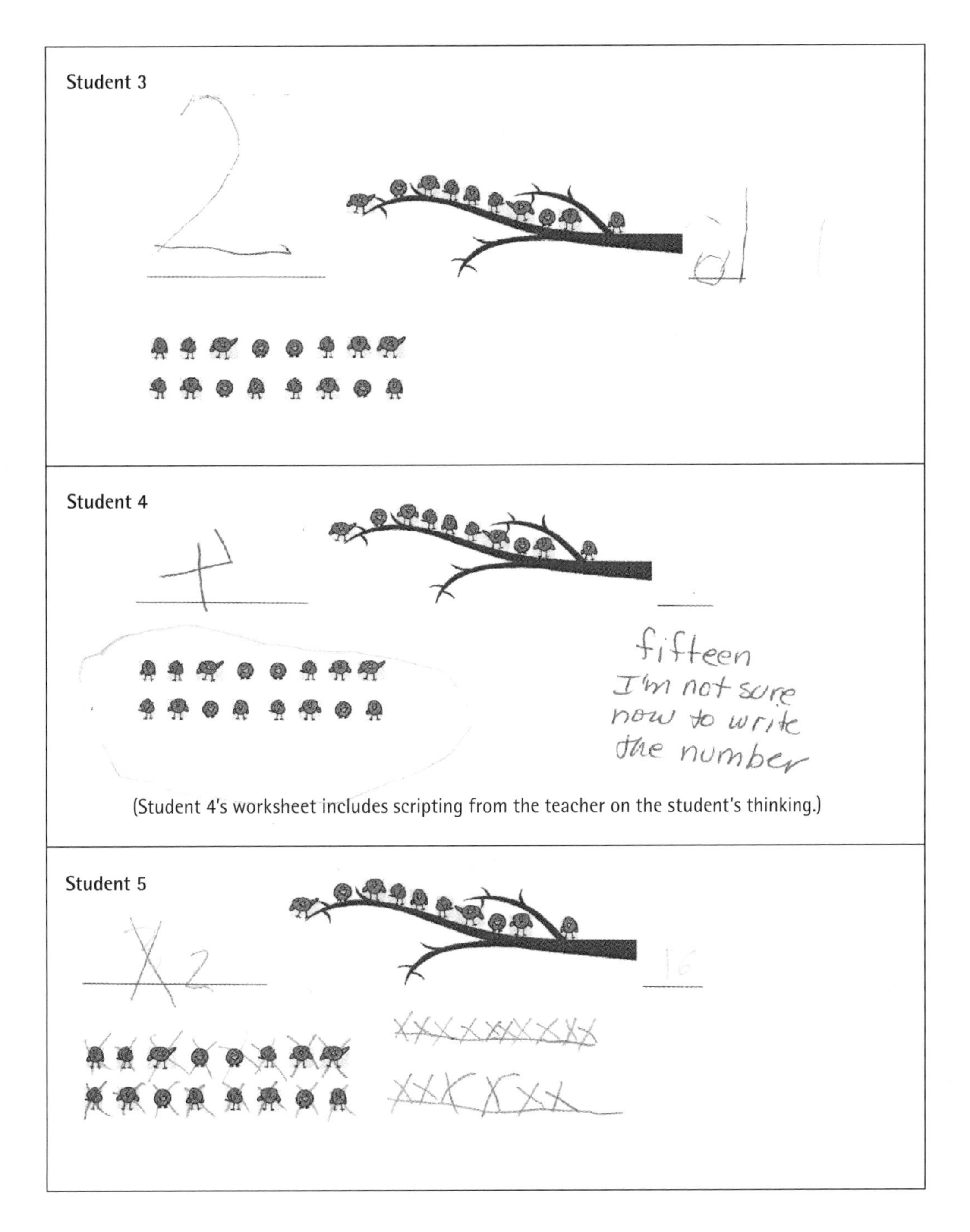

Fig. 4.5. Kindergarteners' work on Birds on a Branch with 16 birds (*continued*)

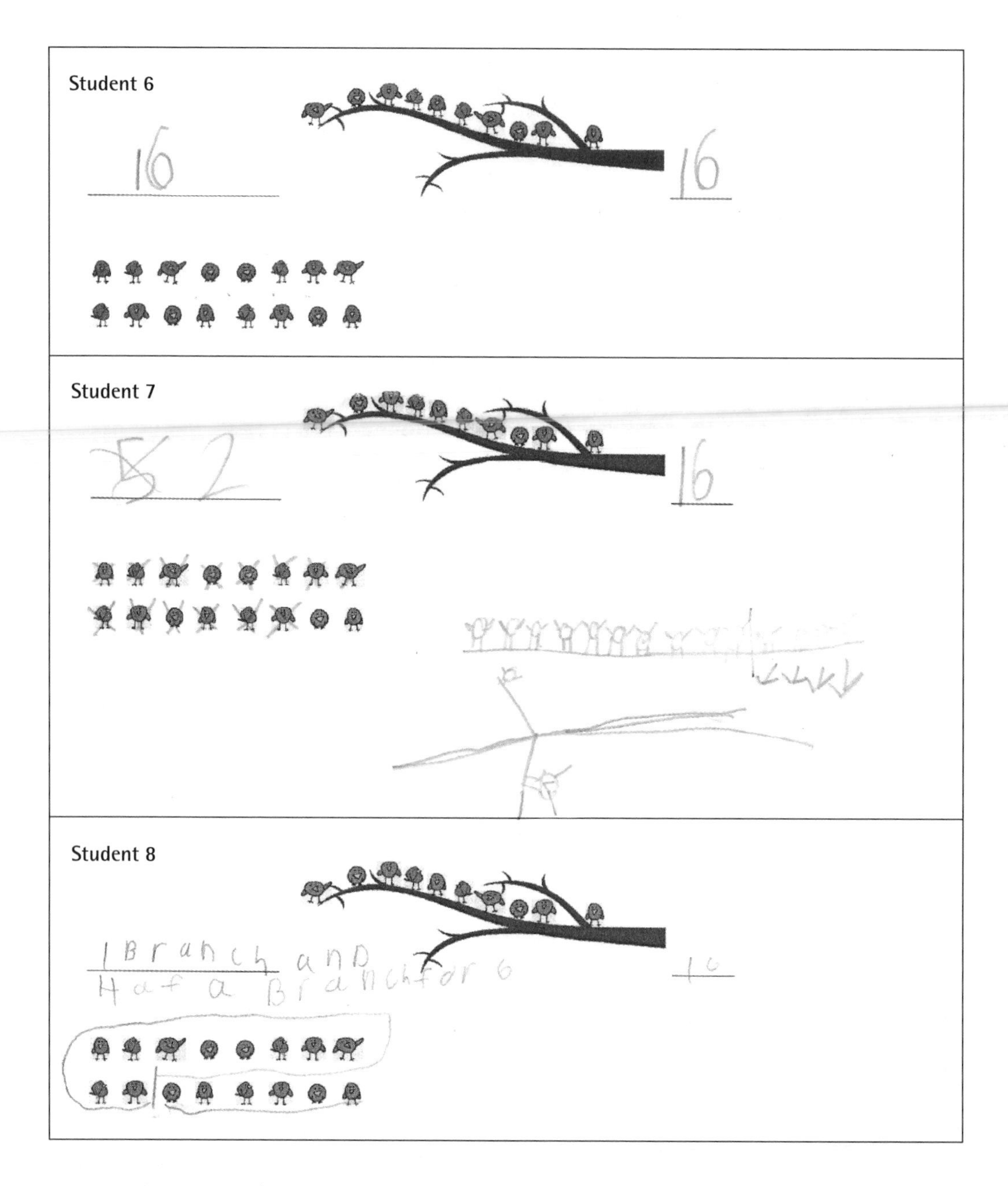

Fig. 4.5. Kindergarteners' work on Birds on a Branch with 16 birds (*continued*)

Students who sketched the situation or crossed off birds as they counted them revealed the organization of their thinking to reach the answer. Work like this is the very kind that students can bring to a document camera to share their thinking with classmates. In this way, students can see other strategies, such as sketching, that help solve problems.

Value of a digit depends on its place or position in a number

Unitizing a quantity into units of ten with units of ones left over is the core of place value. But as noted in these examples, students are challenged to understand that a group of ten items makes "1 ten." This understanding of unitizing; the size of the portion you use when you think about a particular amount is a critical concept in mathematics. As mentioned previously in chapter 1, Lamon (1996) suggested that a case of soda could be thought of as 24 cans (one-units), 2 twelve-packs, or 4 six-packs. The flexibility with units in this soda example, is the same flexibility children need with place value as they interpret 33 as 33 individual units, 3 tens and 3 ones, or even the equivalent value of 2 tens and 13 ones, this last equivalent version being an essential component to regrouping in subtraction. As you can appreciate, unitizing is the underlying framework of place value, and place value is the underlying structure of our system of numeration. These structures are introduced in kindergarten. In later grades, as students reason with very large numbers and decimals, these structures expand their utility. Students who understand place value can identify the creation of a group (e.g., a bundle or a stack of cubes) composed of a particular unit; units established within our base-ten system as powers of ten. The task we examine next focuses on the idea of "ten as 1" and asks students to make an explicit link between the written form of a named amount to the actual concrete quantity.

The Connecting Digits to Values task below connects to the Common Core State Standards (NGA 2010) at a variety of grade levels as kindergarteners work with numbers 11 to 19, first graders recognize that the two digits of a number, such as 53, represent the amounts of tens and ones, and second graders develop the same understandings with the addition of the hundreds in three-digit numbers. There are also connections to the standards for mathematical practices—particularly in practice 7, look for and make use of structure. Although this task involves counting, accurate counting is not the goal. In fact, at this point, students should be accurate counters. Often referred to as the Digit Correspondence Task (Kamii 1985; Ross 1989; 2002), this task has been widely used in the study of assessing students' place-value development. Students are asked to give meaning to a two-digit number through the use of the concrete objects they just counted to match the digits in the numeral. Before presenting the task, see Reflect 4.2 to consider how your students might react to the task. Here is an adaptation of the original task to try with your students.

Reflect 4.2

What do you anticipate the most common student answers might be?

What will be some of the common errors and misconceptions your students may exhibit?

Task: Connecting Digits to Values

This important task helps pinpoint how much students understand about place value.

Do not give clues and do not state the numbers "four" or "two/twenty" as that will influence student thinking. Instead just say "this part of your number."

Take out 24 of the same-size counters.

Ask the student to count the counters.

Ask the student write down on paper the number that tells how many he or she counted. The student should write down "24."

Circle the 4 in 24, and ask the student, "Does this part of your number have anything to do with how many counters there are? Can you show me with the counters?" (CAUTION: do not use the word "four" or say "twenty-four.")

Now circle the 2 (if you think it might be helpful, use two different-colored pens to make your circles) and repeat these questions exactly: "Does this part of your number have anything to do with how many counters there are? Can you show me with the counters?" Again, do not use the words "twenty," "two," or "two tens."

Here are some common misconceptions and naïve understandings that you might expect to observe.

- Difficulty in fully understanding place value and treating the individual digits in the total number all as units might emerge because of previous instruction for adding or subtracting two-digit numbers. For example, in adding 25 + 42, instead of adding twenty and forty (in the tens column), students were encouraged to add "two" and "four," further complicating understanding the concept of place value (Kamii, Lewis, and Livingston 1993).

- Many students (even at fourth and fifth grade) do not fully understand, for example, how in the addition above the single digit number 2 in the tens column represents twenty (Ross 1989). Not surprisingly, a student who lacks this understanding has little likelihood of carrying out two- or three-digit addition and subtraction successfully and meaningfully.

On the basis of her research, Ross (1989; 2002) identified common student responses to the Connecting Digits to Values task that led her to establish five distinct levels of understanding of place value. These range from emerging ideas (level 1) to full understanding (level 5):

Level 1: Single numeral. The student at the single numeral level counts correctly, and accurately writes 24, but views it as a single numeral. When asked about the individual digits circled (4, and later 2), the student cannot say what is meant, suggesting that the digits have no meaning related to the counters.

Level 2: Position names. The student correctly names the 4 as the ones and the 2 as the tens, but when asked to show the relationship to the counters, the student makes no connections between the individual digits of the number and the counters.

Level 3: Face value. This response represents a common level of performance, and one where some students can get stuck. Here, the student accurately matches the circled 4 with four counters, but then matches the circled number 2 with two counters. When asked about the other counters that remain, the child is often unable to say what the other "leftover" counters represent.

Level 4: Transition to place value. At this level, the circled 4 is matched with four counters to represent the ones and the 2 is matched with the remaining collection of twenty counters as a group. In other words, the student points to the rest of the counters and says that is what is connected to the 2. No mention is made of the match to the remaining collection as 2 groups of ten.

Level 5: Full understanding. In this highest level, the 4 is associated with four single counters and the 2 is explicitly stated as being 2 groups of ten counters (taken as a single cluster or even by specific movement of the remaining counters into 2 groups of ten).

To better comprehend these levels of understanding, Reflect 4.3 presents the cases of three students for you to consider and evaluate.

Reflect 4.3

Here are three students' responses to a version of the Connecting Digits to Values Task. At which level of understanding do you think they have reached? What might be a next instructional step for each?

Leo. The teacher first asks Leo to count out a collection of 20 counters from a bucket of counters. Leo counts 18 counters without an organized approach and when the teacher says how many do you have, he then says, "18." The teacher follows up with "But I asked you to count out 20." Leo quickly adds 2 more counters to the group. The teacher then requests, "Take out 4 more counters and tell me how many you have now." Leo counts 4 more and counts on with his fingers and says, "24." "Can you write that number on the paper?" asks the teacher. Leo writes 24. The teacher then circles the 4 and asks "Does this part of your number have anything to do with how many counters there are? Can you show me with the counters?" Leo takes out 4 counters. The teacher then circles the 2 and says, "Show me this part of your number with your counters." Leo counts out 2 counters.

Maria. The teacher asks Maria to complete the same counting of 20 counters, and Maria counts out 20 counters, placing them into one long row of 20. The teacher asks "How many counters do you have?" Maria says "20." The teacher then asks, "Can you add 4 more counters to your pile?" Maria adds 4 more to her row, and says unprompted, "Now I have 24." The teacher says, "You do, can you write that number?" Maria writes the number 24. The teacher circles the 4 and asks the question described in the task. Maria pulls away the last 4 counters from her line of counters. Then the teacher circles the 2 and asks, "How is this number related to your counters?", and Maria says, "It is the rest of them."

John. John is asked to count out 20 counters; he counts out 20 and places them into two piles of 10 counters each. The teacher then asks, "How many do you have?" John says "20." The teacher then asks John to add 4 more, and he does so by creating another separate pile to the right of the other sets of counters. John says unprompted, "Now I have 24." The teacher says, "You do. Can you write that number?" John writes the number 24. The teacher circles the 4 and asks, "Does this part of your number have anything to do with how many counters you have?" John says "Four is in the ones place," and points to the 4 counters in the small pile. The teacher says as she circles the 2, "Does this part of your number have anything to do with how many counters there are?" John replies, "That is the 2 groups of ten that I counted," and points to both groups of 10 counters.

From administering these three interviews, we observed that the students' responses aligned with the findings from Ross (1989; 2002) and reveal three different levels of understanding of place-value. Leo aligns with level 3 when he matches 4 counters with the 4 and 2 counters with the 2, taking the digits on "face value." Maria's response marks the transition to place value understanding where the 4 is matched with a set of 4 counters and the 2 is linked to the remaining counters as a whole group. The final level, or "full understanding," is where a student, John in this case, indicates that the 2 is correlated with 2 groups of ten counters and the 4 with four single counters.

For students who struggle with the Connecting Digits to Values task, we provide several other follow-up activities that are useful:

Activity: Count in Piles. Students select various items in the classroom or items brought in by the teacher (straws, pencils, books on the reading shelf, etc.), and are asked to strategize how these items can be counted in ways other than by ones. If students don't suggest it, eventually you may want to move them to the suggestion of counting the group by tens.

Activity: Seeking Tens. Prepare small plastic bags that contain counters or other items (craft sticks, toothpicks, pennies, etc.). Students then count the items in the bag, record it as a numeral, record it as a number word, and mark it on a ten and singles sheet or a ten-frame sheet. Bags are traded, and new counts are taken and recorded. As they get the idea, you can extend the task and ask students to make estimates first.

Students should progress from a unitary conception where they only count by ones (all individual units) to a sequence-by-tens approach (ten, twenty, thirty, forty, forty-one, forty-two), and finally to a separate tens-and-ones approach (3 tens and 4 ones—34) (Fuson 1990). If your students are still interpreting double-digit amounts at any of the levels of understanding from 1 to 3, there is more instructional work to be done with groupable models of ten. Materials such as connecting cubes, coffee stirrers or craft sticks, or a sheet of double ten-frames can be used to make tens and reinforce the counting of the tens in the process of counting out the amounts in two-digit numbers. After students have had experiences building groups of ten and recognizing these groups as a unit of ten, they can advance to the use of semi-concrete representations of dot cards of two-digit numbers or line-and-dot illustrations that record the base-ten materials through sketches. The drawings of lines and dots to represent tens and ones can also be used to play a matching game where children pair numerals to the corresponding line-and-dot illustrations (see fig. 4.6). These connections and relationships build students' understanding of place value and move them away from thinking that multidigit numbers are a collection of separate single-digit numbers.

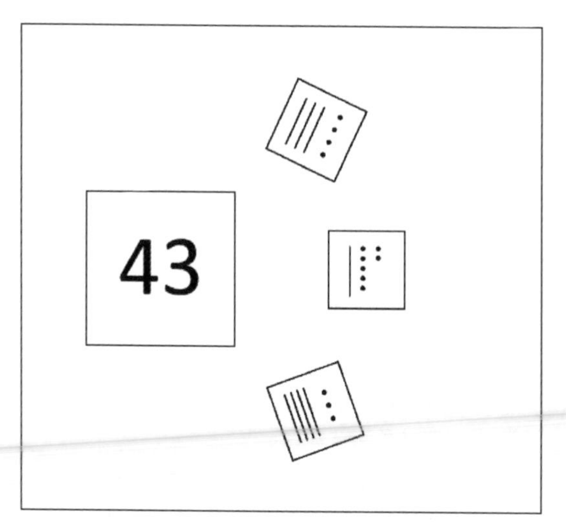

Fig. 4.6. Matching cards for Line-and-Dot game

As an aside, it is critical to foster students' use of sketches and drawings by sharing these common methods of recording results in a semi-concrete format—such as line-and-dot illustrations. Without such guidance, many students can spend an inordinate amount of time trying to draw elaborate illustrations of base-ten materials. Time is better spent thinking about the mathematical relationships.

Also, be vigilant about avoiding the use of nonproportional base-ten materials. Instead of having models where 1 ten is a match for 10 ones, nonproportional models do not accurately represent the physical proportional relationship; for example, money (a dime does not physically equal ten pennies), chips (a blue chip does not physically equal ten white chips), or a pocket chart where one craft stick can be a one in the Ones pocket or a ten in the Tens pocket, and so on. (However, there are proportional relationships that can be shown using pocket charts where one craft stick represents a one in the Ones pocket, and a bundle of ten sticks represents a ten in the Tens pocket.) Additionally, commercially available base-ten materials should be reinforced to their link to place value units by calling them ones, tens, and hundreds rather than by their shapes, such as naming them rods or flats.

Task: Changing by Ten

The goal is to move students from counting by ones to adding ten.

The student is first given a problem, with base-ten materials available, where they are asked to–

- Show 14. Then once shown, change 14 to 16.
- Show 7. Then once shown, change 7 to 17.
- Show 12. Then once shown, change 12 to 22.
- Show 15. Then once shown, change 15 to 25.

Students start by adding ones to the 14 to create 16, but at some point they should stop counting by ones and instead just add a ten. That is the important milestone to assess with this task.

Task: Number Stack Cards

Students practice linking numerals to quantities in purposeful ways.

In modeling these various quantities what is critical is to link the counts of the items to the corresponding numerals and number words. Use the small Number Stack Cards (see fig. 4.7 and Appendix 3) that have the numerals for the decade number; for example, the 0 on the 20 card (showing the two digits) can be covered by cards with all the possible ones digits–in this case 4–to show how the numeral 24 is actually the 20 with the number 4 stacked on top, allowing the individual components to remain but be combined to make the 24. Link this model to the use of a place-value mat by putting the number cards under the values, and then overlaying the cards with a right justification. The critical idea is to have students link numerals to quantities in meaningful ways.

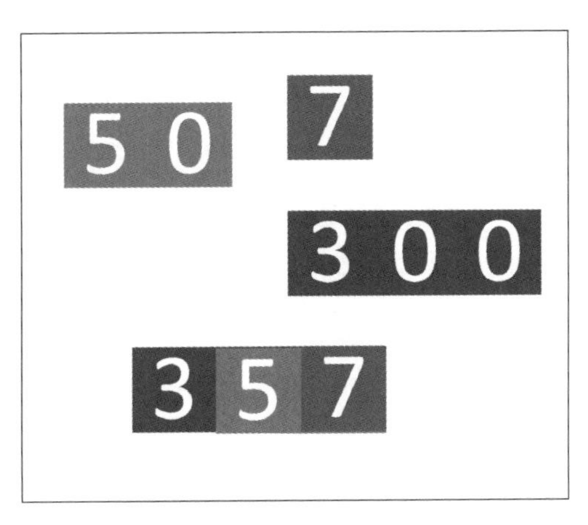

Fig. 4.7. Number Stack cards

Task: Mixed-Up Places

Students sort out numbers that are described out of place-value order.

The idea of this task is to give students a written description of a number indicating how many ones, tens, and, in some cases, hundreds it has—but not in the traditional order of the place-value positions (see Appendix 3 for a sheet with several options). Students' initial tendency might be to assume the number positions are in the correct order and name the number from the numerals left to right. To identify common confusions and emerging ideas, let's look at the following examples of two students' work (see figs. 4.8 and 4.9).

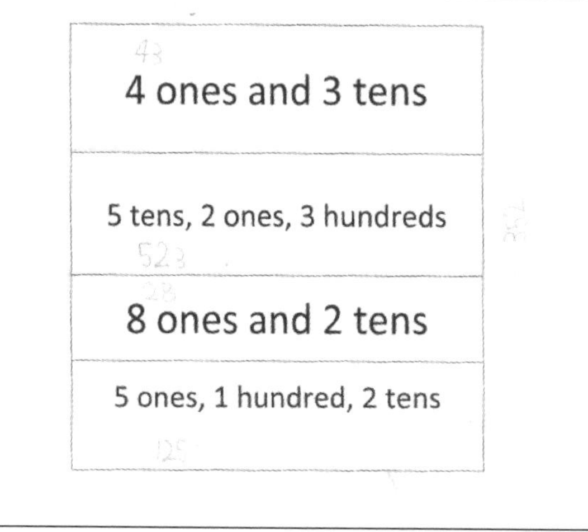

Fig. 4.8. Rana's work

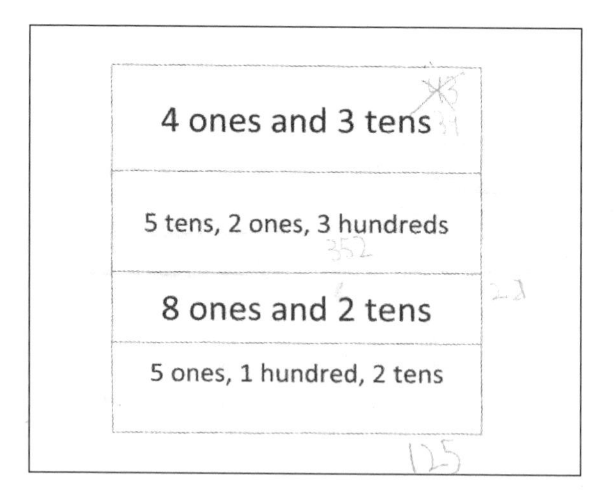

Fig. 4.9. Dora's work

Rana's responses show her deepening understanding from following the order left to right to a shift in the second example for which she generates two answers. Then she moves to the correct numbers.

Dora too started off by writing 43, but then we offered base-ten materials and a place-value mat. By creating the number with the tens and ones pieces (following the words for directions) she rethought her original answer and crossed it out to be replaced by 34. She also used the materials for the second problem, but then completed the rest without the use of concrete materials.

Summary: Learners, Curriculum, Instruction, and Assessment

To effectively teach the mathematical ideas presented in this chapter, teachers must have knowledge of the four components (learner, curriculum, instructional strategies, and assessment) presented in the Introduction. In the following sections we summarize some key ideas for each of these components.

Knowledge of learners

When students see numbers as just a series of digits where, for example, they believe the 9 in the ones place of 2,149 has the greatest value, we recognize that they do not yet grasp the structure of our number system. We can improve their place-value thinking by demonstrating the three-way connection between oral number names, written numerals, and the standard and equivalent groupings with materials or drawings that represent quantities. By connecting these ideas, we can highlight the patterns and structure of our number system and illuminate the values of the numbers they will be operating on in the coming years.

Knowledge of curriculum

The curriculum in the U.S. pertaining to our conventions for stating numbers accentuates cross-cultural differences, as described above. In an effort to have U.S. students reach greater success in place-value understanding, some educators suggest that we have our students change the names of the numbers to "two-ten-one" for 21, for example. This approach is shortsighted and disassociates children from the world in which they live.

Knowledge of instructional strategies

The use of concrete models that connect proportionally to the values in the place-value system is a critical starting point to shifting students from a unitary approach to thinking about our base-ten system. Of course, the numeral as a symbolic connection and the corresponding number names should be shared simultaneously. This relational understanding of place value integrates these three components to ensure the realization by students that numbers are not just a series of single digits, but are, in fact, in a multiplicative relationship. They become aware that each place is ten times the value of the place to the right of it, eventually learning that each place is also one-tenth of the value of the place to the left of it. By using proportional materials, students can explain the relationship and use materials to demonstrate their thinking.

Knowledge of assessment

The assessment of place-value knowledge is essential; it can never be assumed. Students will not be able to add, subtract, multiply, or divide whole numbers without this knowledge. Neither will they be able to interpret reasonableness of answers or carry out accurate computational estimation. By implementing the various tasks in this chapter in the form of diagnostic interviews, students' knowledge and strengths can be converted into tailored instruction.

Conclusion

Not only do the foundational ideas of place value extend over the elementary years but they also remain the key ingredients to future mathematical endeavors. The standards of every state emphasize place-value concepts as essential domains for study and as nonnegotiable concepts that must be learned to effectively proceed in the mathematics curriculum. It is important and necessary that enough time be spent so that students attain a thorough and deep understanding of this area of study because without it, students will be relegated to struggle with the concepts awaiting them in future learning.

into practice

Chapter 5 Looking Ahead with Number and Numeration

Chapter 5 highlights how the big ideas and essential understandings for number and numeration described and demonstrated in chapters 1 to 4 align with mathematics concepts that students develop after grade 2. This discussion demonstrates how important it is for students in pre-K–grade 2 to develop a deep understanding of the essential concepts and skills that serve as a foundation for subsequent learning.

Extending Knowledge of Number and Numeration in Grades 3–5

The understanding of number and numeration that is developed in pre-K–grade 2 lays the foundation for the number ideas that will be developed in grades 3–5. There are two primary aspects of number and numeration that provide this foundation: (1) concept of unit and (2) counting structures. In the sections below, we provide further connection to the big ideas and essential understandings and the Common Core State Standards for Mathematics (CCSSM; National Governors Association Center for Best Practices and Council of Chief State School Officers [NGA Center and CCSSO 2010]).

Building on student understanding of concept of unit

A shift that occurs in grades 3–5 is the extension of investigating and operating solely on whole numbers to include rational numbers. Sometimes students believe there is little or no connection between whole numbers and rational numbers, such as fractions, for example. Fractions look very different in their form; additionally, the way in which we compute with fractions seems to mask the relationship between the two number systems. However, students' concept of unit is a significant connection between the two number systems. The whole, or unit, from which fractions are constructed is the basis for the conceptual and procedural aspects that form the foundation of rational numbers. As with whole number comparisons, in order to compare fractions, the unit or whole must be the same. Without that requisite condition, the comparability of fractions is lost, and we cannot determine whether $\frac{1}{2}$ is greater than or less than $\frac{3}{4}$.

Building on student understanding of place value

As students progress from pre-K–grade 2 to grades 3–5, they also utilize the concept of unit to understand the structure of the rational number system. The concept of unit is an important building block that forms the foundational aspects of place value. In pre-K–grade 2, students explore how the unit (ones) are iterated ten times to make the next place value (tens), and that the new unit iterated ten times makes the next place value (hundreds), and so on. The relationship that each place-value position is ten times greater than the one to the right of it is crucial in understanding the relationships represented within and across numbers.

The relationship expressed in place value with whole numbers is extended in grades 3–5: In third grade, place value understandings are put into use when students carry out multidigit computations and round whole numbers to the nearest ten or hundred. Decimals are introduced in grade 4, and their study continued in grade 5 as recommended by the Common Core State Standards for Mathematics.

Common Core State Standards for Mathematics

Number and Operations in Base-Ten

Grades 3–5

- *3.NBT.1.* Use place value understanding to round whole numbers to the nearest 10 or 100.
- *3.NBT.2.* Fluently add and subtract within 1000 using strategies and algorithms based on place value, properties of operations, and/or the relationship between addition and subtraction.
- *4.NF.6.* Use decimal notation for fractions with denominators 10 or 100. For example, rewrite 0.62 as $\frac{62}{100}$; describe a length as 0.62 meters; locate 0.62 on a number line diagram.
- *4.NF.7.* Compare two decimals to hundredths by reasoning about their size. Recognize that comparisons are valid only when the two decimals refer to the same whole. Record the results of comparisons with the symbols >, =, or <, and justify the conclusions, e.g., by using a visual model.
- *5.NBT.2.* Explain patterns in the number of zeros of the product when multiplying a number by powers of 10, and explain patterns in the placement of the decimal point when a decimal is multiplied or divided by a power of 10. Use whole-number exponents to denote powers of 10.
- *5.NBT.3.* Read, write, and compare decimals to thousandths.
- *5.NBT.4.* Use place value understanding to round decimals to any place.

The relationship between and among place-value positions that is established in the early grades plays an important role as students move forward in their study of mathematics. The generalization that the place-value position to the left of another place-value position is ten times greater, and conversely the place-value position to the right of another position is one-tenth of that position's value, is a major understanding that students need in order to understand decimals conceptually and procedurally.

With regard to the procedural aspects, the place-value understandings support the algorithms that are developed in grades 3–5. The traditional or standard algorithm as well as other algorithms, such as partial sums, is more powerful when students can determine the reasonableness of answers using their understandings of place value. Predicting the magnitude of the answer before doing the computation, and then assessing the reasonableness of the answer after performing the computation provide opportunities for students to self-check and monitor their work.

Summary

Number development in pre-K–grade 2 is foundational to the more sophisticated and complex number concepts and skills that arise in later grades. Without the prerequisite understandings that are developed in these early grades, students are likely to think of a number in a superficial way and never really develop the deeper understandings that promote number flexibility. The ability to think flexibly about number builds students' confidence and proficiency.

Appendix 1

The Big Ideas and Essential Understandings for Number and Numeration

This book focuses on the big ideas and essential understandings that are identified and discussed in *Developing Essential Understanding of Number and Numeration for Teaching Mathematics in Prekindergarten–Grade 2* (Dougherty, Flores, Louis, Sophian, and Zbiek 2010). For the reader's convenience, the complete list of the big ideas and essential understandings in that book is reproduced below.

Big Idea 1. Number is an extension of more basic ideas about relationships between quantities.

Essential Understanding 1*a*. Quantities can be compared without assigning numerical values to them.

Essential Understanding 1*b*. Physical objects are not in themselves quantities. All quantitative comparisons involve selecting part icular attributes of objects or materials to compare.

Essential Understanding 1*c*. The relation between one quantity and another quantity can be an equality or inequality relation.

Essential Understanding 1*d*. Two important properties of equality and order relations are conservation and transitivity.

Essential Understanding 1*e*. The equality relation between two quantities remains unchanged when one or both quantities are decomposed into parts and when one of the quantities is combined with another quantity to form a larger quantity.

Big Idea 2. The selection of a unit makes it possible to use numbers in comparing quantities.

Essential Understanding 2*a*. Using numbers to describe relationships between or among quantities depends on identifying a unit.

Essential Understanding 2*b*. The size of a unit determines the number of times that it must be iterated to count or measure a quantity.

Essential Understanding 2*c*. Quantities represented by numbers can be decomposed (or composed) into part-whole relationships.

Big Idea 3. Meaningful counting integrates different aspects of number and sets, such as sequence, order, one-to-one correspondence, ordinality, and cardinality.

Essential Understanding 3*a*. The number-word sequence, combined with the order inherent in the natural numbers, can be used as a foundation for counting.

Essential Understanding 3*b*. Counting includes one-to-one correspondence, regardless of the kind of objects in the set and the order in which they are counted.

Essential Understanding 3*c*. Counting includes cardinality and ordinality of sets of objects.

Essential Understanding 3*d*. Counting strategies are based on order and hierarchical inclusion of numbers.

Big Idea 4. Numbers are abstract concepts.

Essential Understanding 4*a*. Patterns in the number-word sequence provide a foundation for the abstract number concept.

Essential Understanding 4*b*. The number sequence is infinite.

Essential Understanding 4*c*. Number symbols are representations of abstract mental objects.

Big Idea 5. A base-ten positional number system is an efficient way to represent numbers in writing.

Essential Understanding 5*a*. Ten different digits can be used and sequenced to express any whole number.

Essential Understanding 5*b*. Our base-ten number system allows forming a new place-value unit by grouping ten of the previous place-value units, and this process can be iterated to obtain larger and larger place-value units.

Essential Understanding 5*c*. The value of a digit in a written numeral depends on its place, or position, in a number.

Essential Understanding 5*d*. Inherent in place value are units of different size.

Appendix 2

Resources for Teachers

The following list highlights a few of the many books and articles that are helpful resources for teaching number and numeration in prekindergarten–grade 2.

Books

Battista, Michael T. *Reasoning and Sense Making in the Mathematics Classroom, Pre-K-Grade 2.* Reston, Va.: National Council of Teachers of Mathematics, 2016.

Brosterman, Norman. *Inventing Kindergarten.* New York: Kaleidograph Design, 2014.

Chapin, Suzanne H., and Art Johnson. *Math Matters: Understanding the Math You Teach, Grades K–8.* 2nd ed. Sausalito, Calif.: Math Solutions, 2006.

Copley, Juanita V. *Showcasing Mathematics for the Young Child: Activities for Three-, Four-, and Five-Year Olds.* Reston, Va.: National Council Teachers of Mathematics, 2004.

———. *The Young Child and Mathematics.* 2nd ed. Washington, D.C.: National Association for the Education of Young Children; Reston, Va.: National Council of Teachers of Mathematics, 2010.

Early Math Collaborative, Erikson Institute. *Big Ideas of Early Mathematics: What Teachers of Young Children Need to Know.* Boston: Pearson, 2014.

Fuson, Karen C., Douglas H. Clements, and Sybilla Beckmann. *Focus in Kindergarten: Teaching with Curriculum Focal Points.* Reston, Va.: National Council of Teachers of Mathematics; Washington, D.C.: National Association for the Education of Young Children, 2010.

Keeley, Page, and Cheryl Rose Tobey. *Mathematics Formative Assessment.* Thousand Oaks, Calif.: Corwin; Reston, Va.: National Council of Teachers of Mathematics, 2011.

National Association for the Education of Young Children (NAEYC). *Exploring Math & Science in Preschool.* Washington, D.C.: NAEYC, 2015.

National Research Council. *Mathematics Learning in Early Childhood: Paths toward Excellence and Equity.* Committee on Early Childhood Mathematics, Christopher T. Cross, Taniesha A. Woods, and Heidi Schweingruber, eds. Center for Education, Division of Behavioral and Social Sciences and Education. Washington, D.C.: National Academies Press, 2009.

Pollman, Mary Jo. *Blocks and Beyond: Strengthening Early Math and Science Skills through Spatial Learning.* Baltimore: Paul H. Brookes, 2010.

Sarama, Julie, and Douglas H. Clements. *Early Childhood Mathematics Education Research: Learning Trajectories for Young Children.* New York: Routledge Taylor & Francis, 2009.

Journal Articles

Betts, Paul. "Counting on Using a Number Game." *Teaching Children Mathematics* 21 (March 2015): 430–436.

Russo, James. "Skip-counting Battle." *Teaching Children Mathematics* 22 (April 2016): 512.

Schwerdtfeger, Julie and Angela Chan. "Counting Collections." *Teaching Children Mathematics* 13 (March 2007): 356–361.

Sci, Eve, Kristen Sendrowski Kircher, and Heather Shook. "Complex Counting in Kindergarten." *Teaching Children Mathematics* 22 (March 2016): 434–441.

Appendix 3
Tasks

This book examines rich tasks that have been used in the classroom to bring to the surface students' understandings and misunderstandings about number and numeration. A sampling of these tasks is offered here, in the order in which they appear in the book. At More4U, Appendix 3 includes these tasks, some with templates for classroom use.

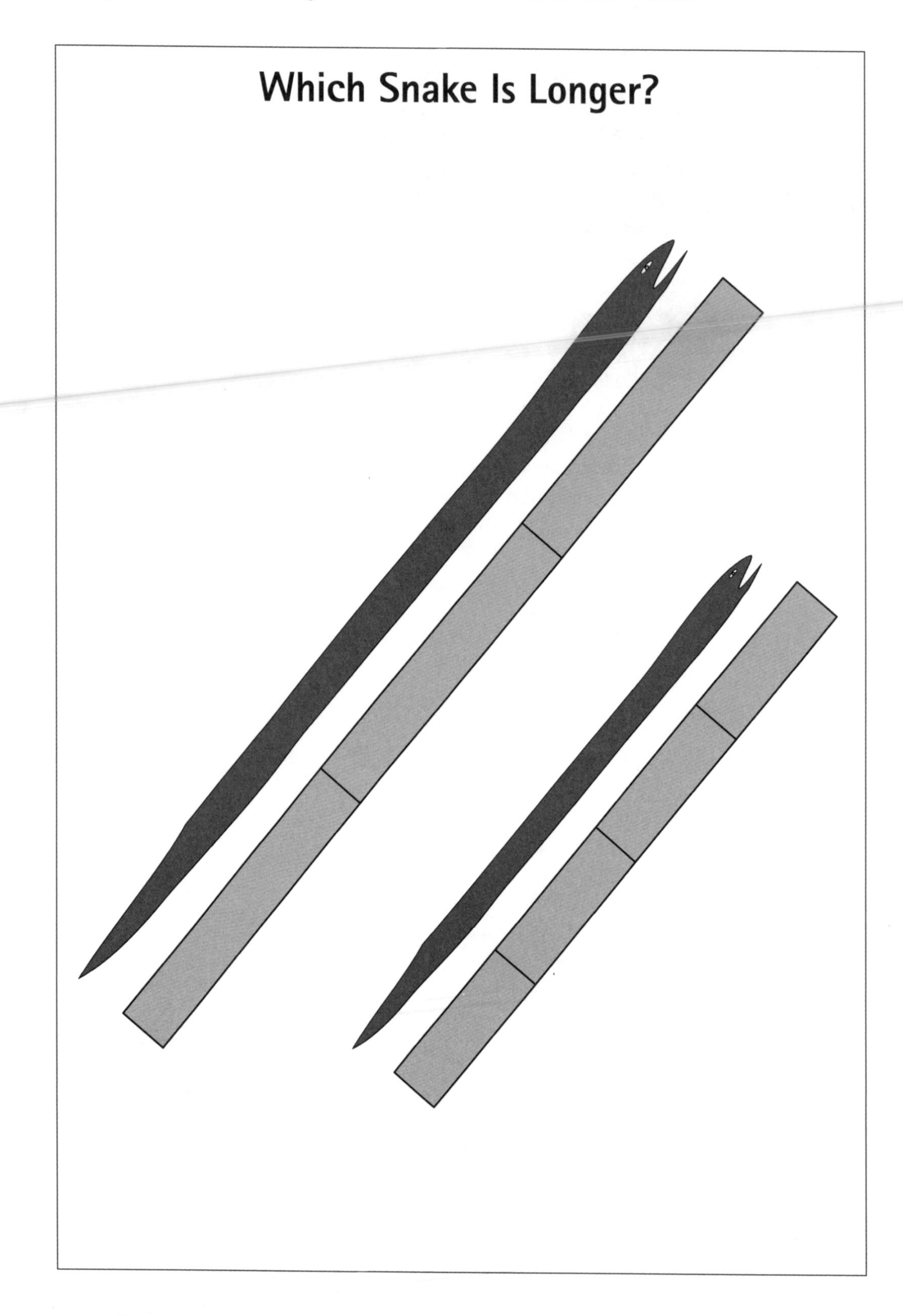
Which Snake Is Longer?

Cookie Comparison Tasks

Show students the following pairs of plates of cookies. Say Hungry Puppet wants to eat the most cookies. Which plate has the most cookies? Circle the choice that the student makes.

Trial 1

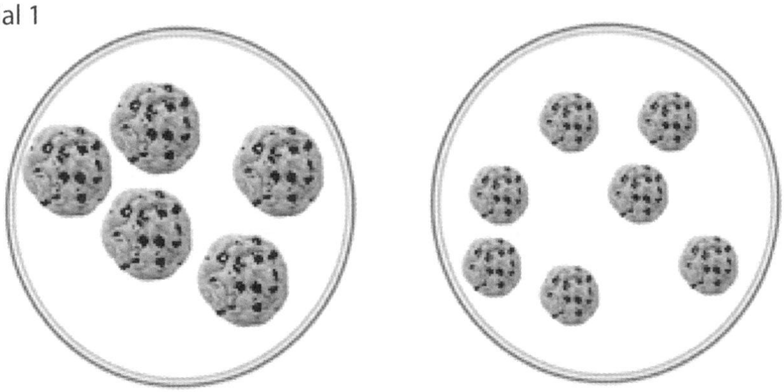

Trial 2

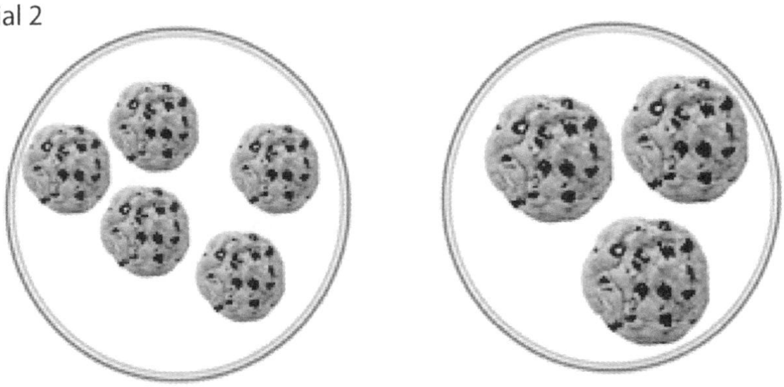

Trial 3

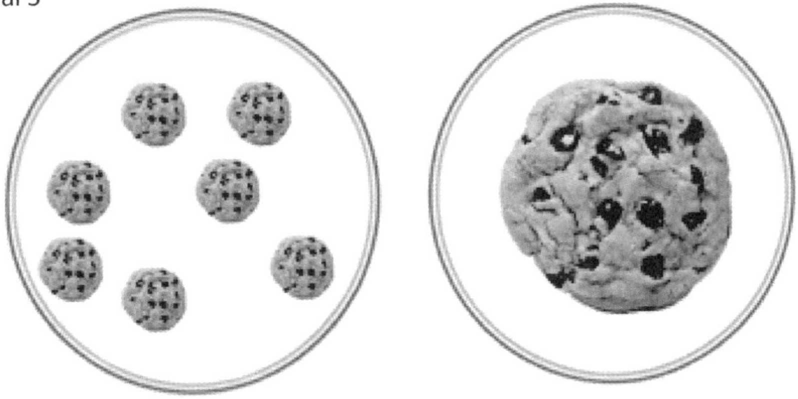

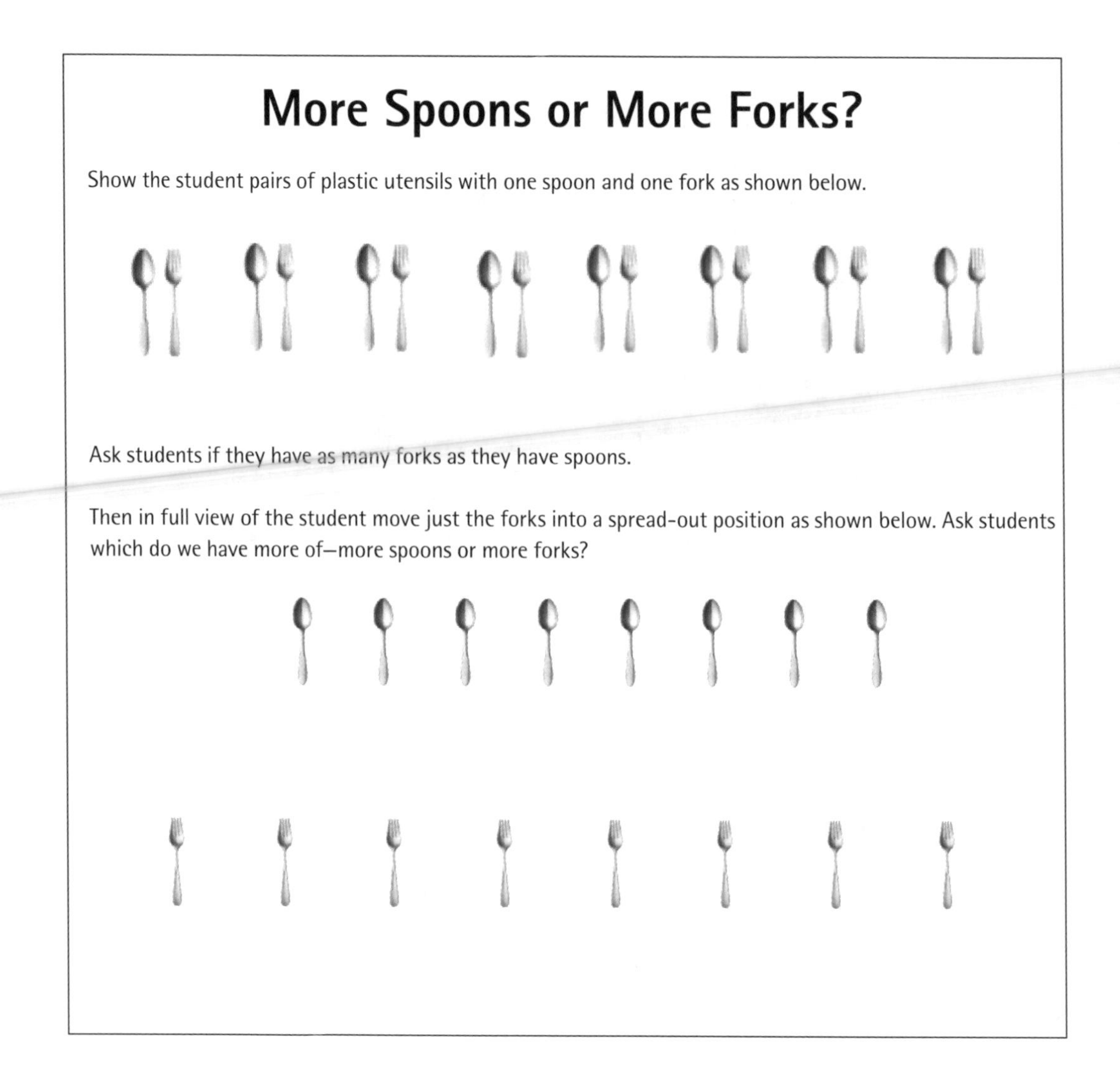

More Spoons or More Forks?

Show the student pairs of plastic utensils with one spoon and one fork as shown below.

Ask students if they have as many forks as they have spoons.

Then in full view of the student move just the forks into a spread-out position as shown below. Ask students which do we have more of—more spoons or more forks?

Peanut Task, Part I

Show students the following image. Point to the first row with your finger moving across the row. Point to the second row with your finger moving across the row. Ask the student, "Which row has more peanuts?"

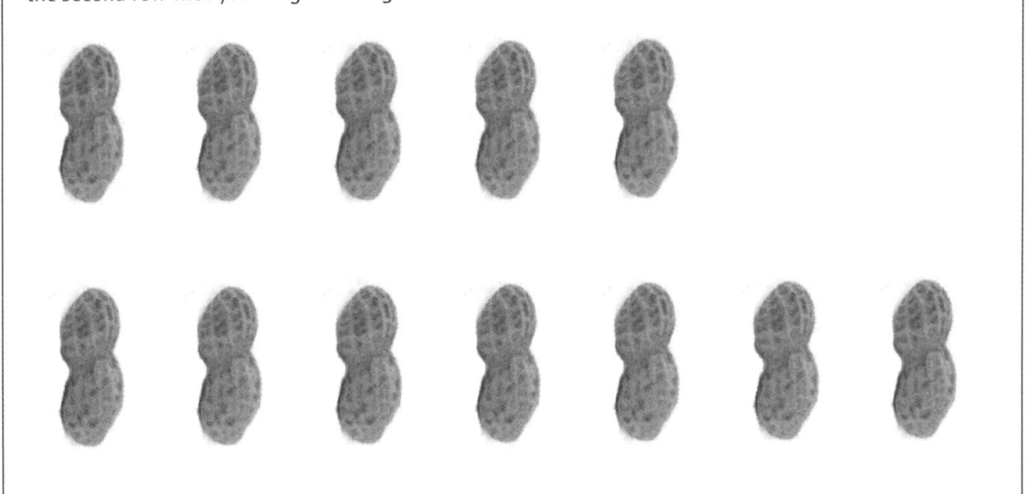

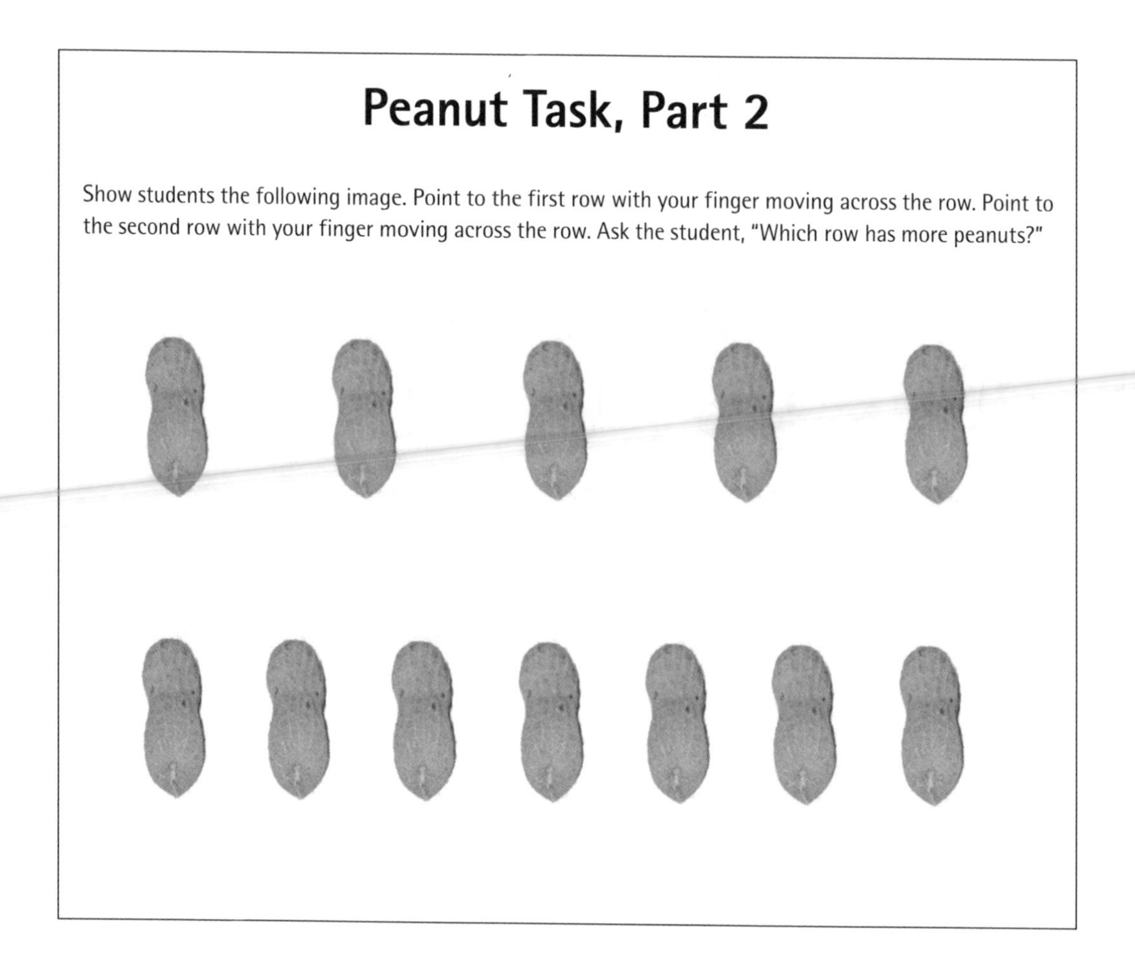

Peanut Task, Part 2

Show students the following image. Point to the first row with your finger moving across the row. Point to the second row with your finger moving across the row. Ask the student, "Which row has more peanuts?"

Crossing Over Cards

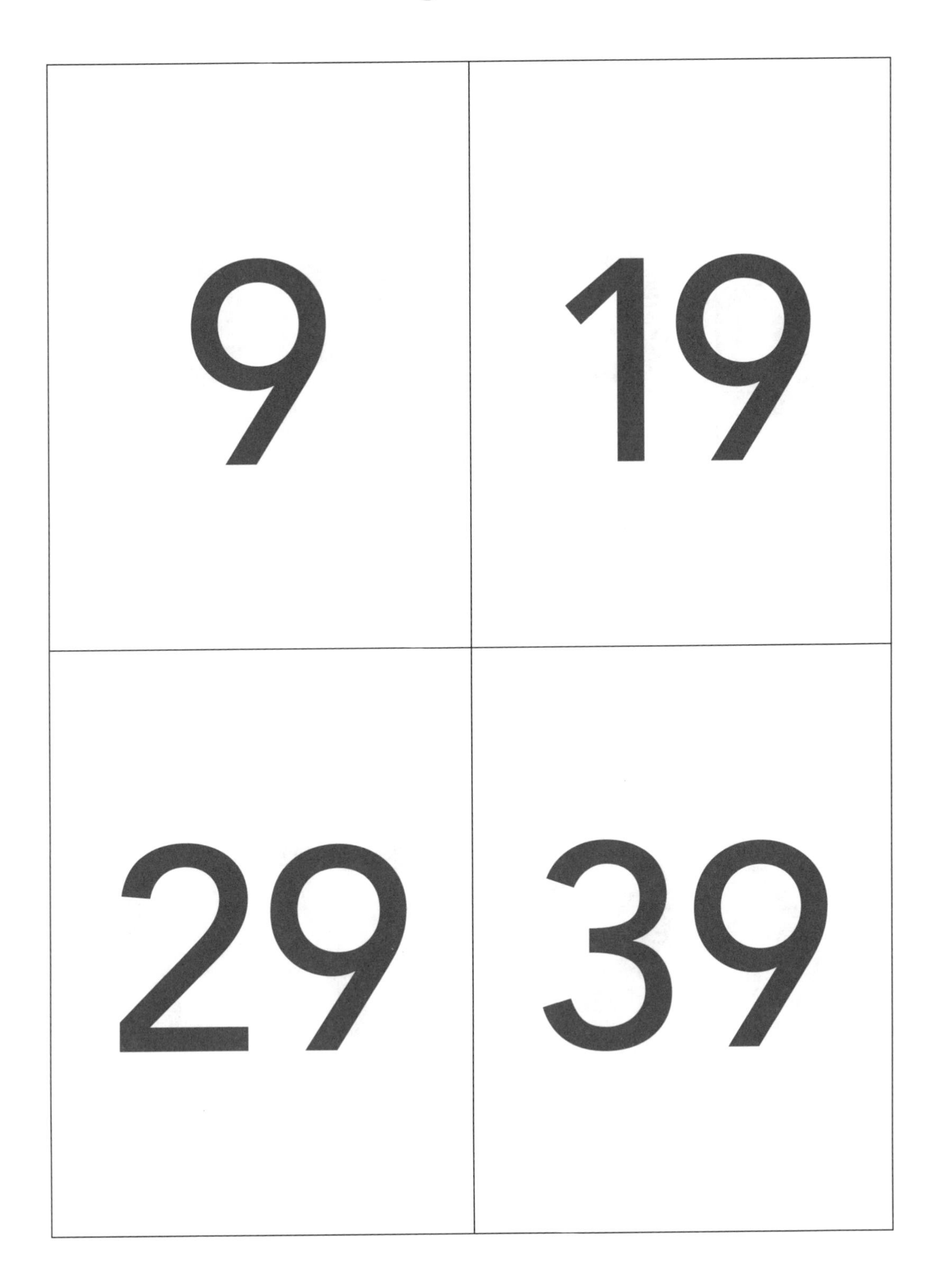

Crossing Over Cards

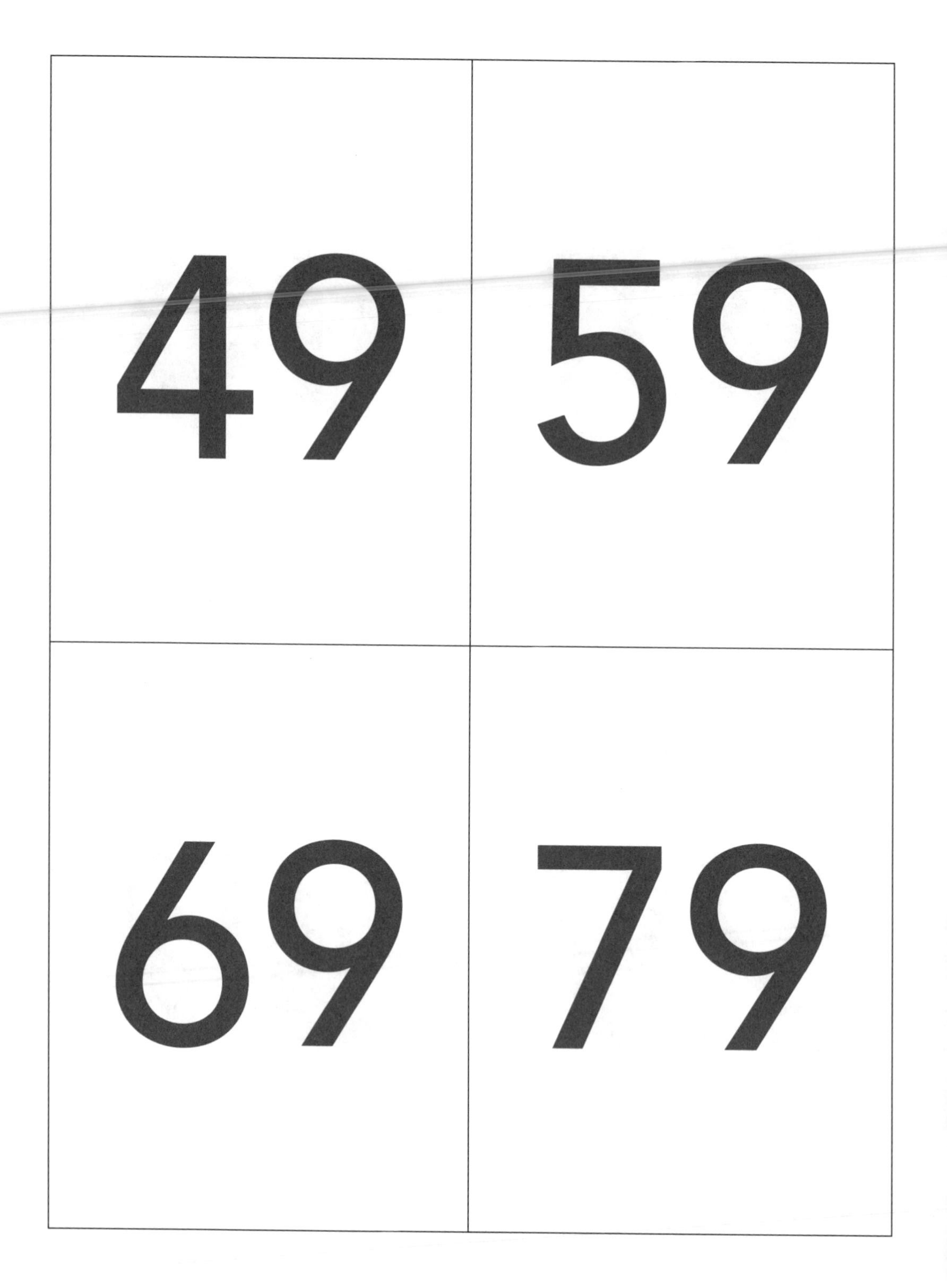

Crossing Over Cards

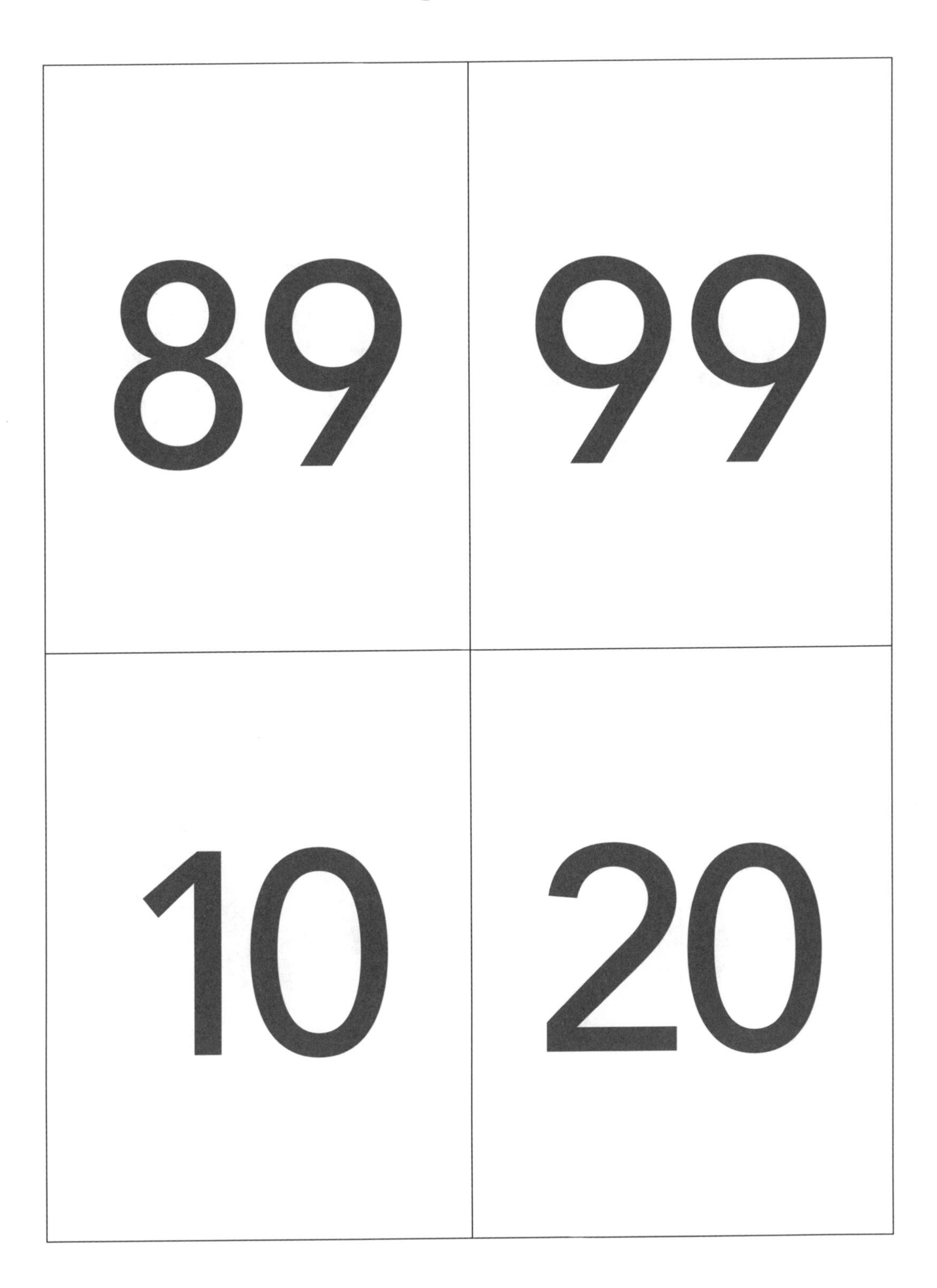

Crossing Over Cards

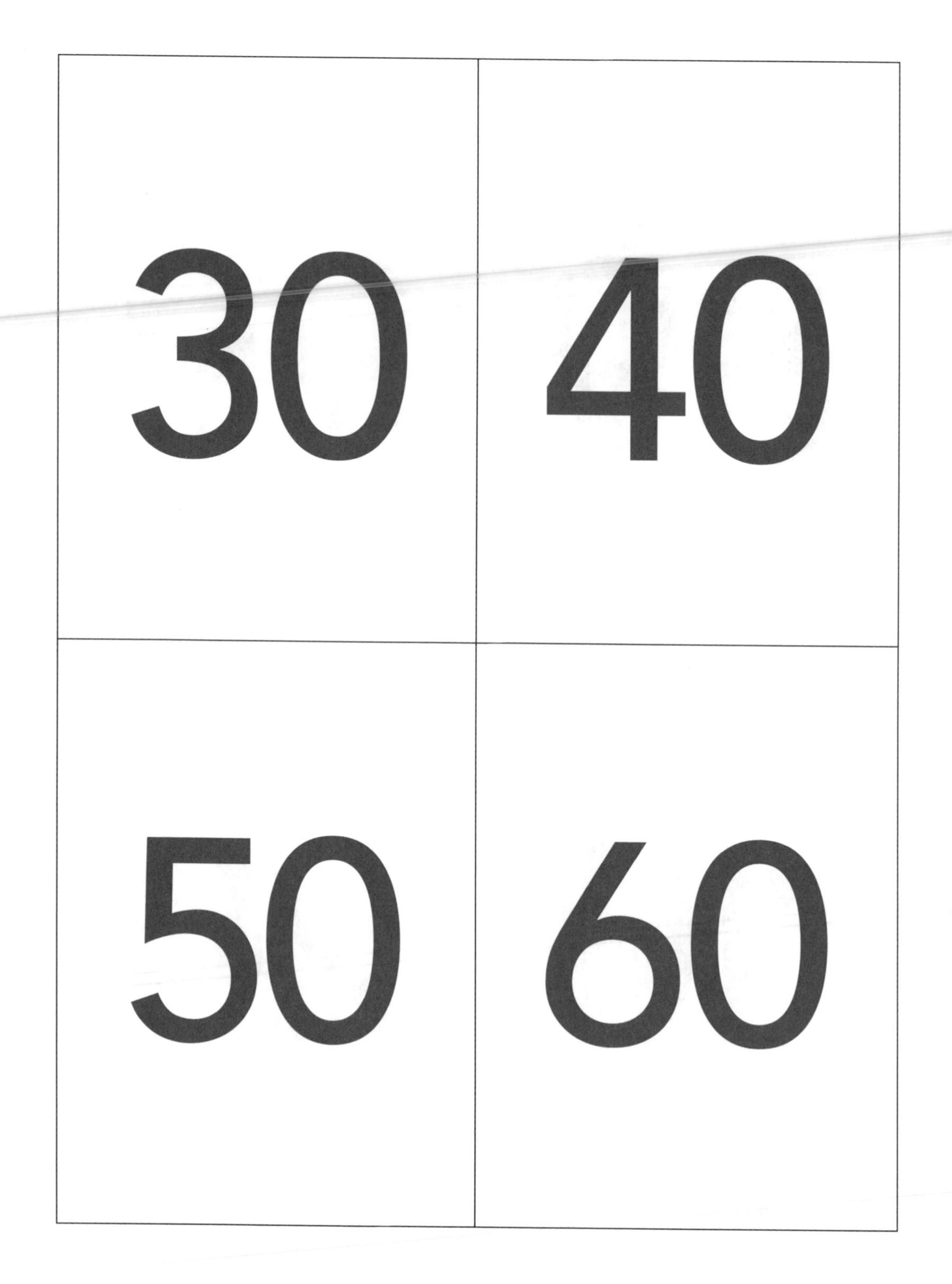

Crossing Over Cards

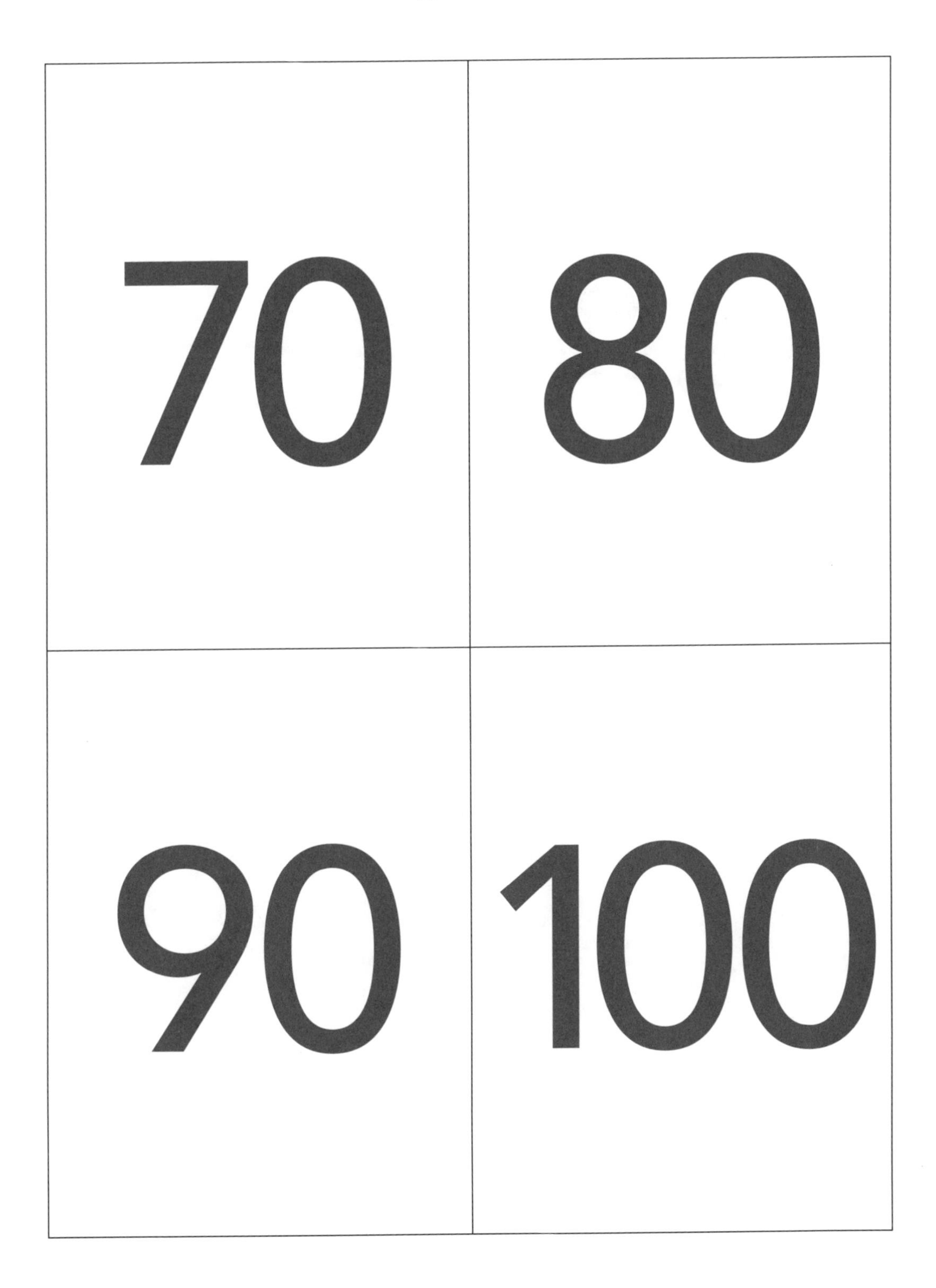

Crossing Over Cards

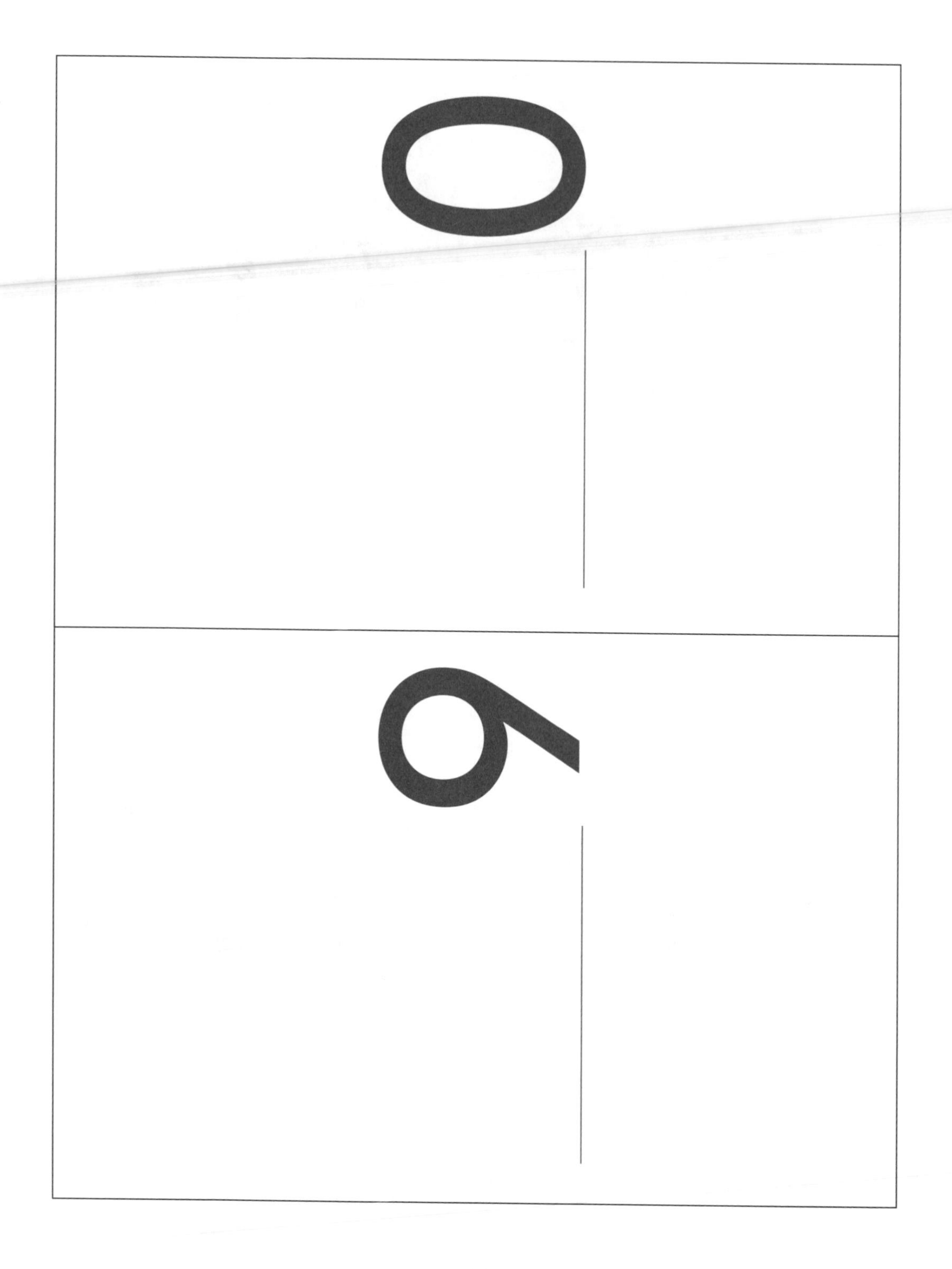

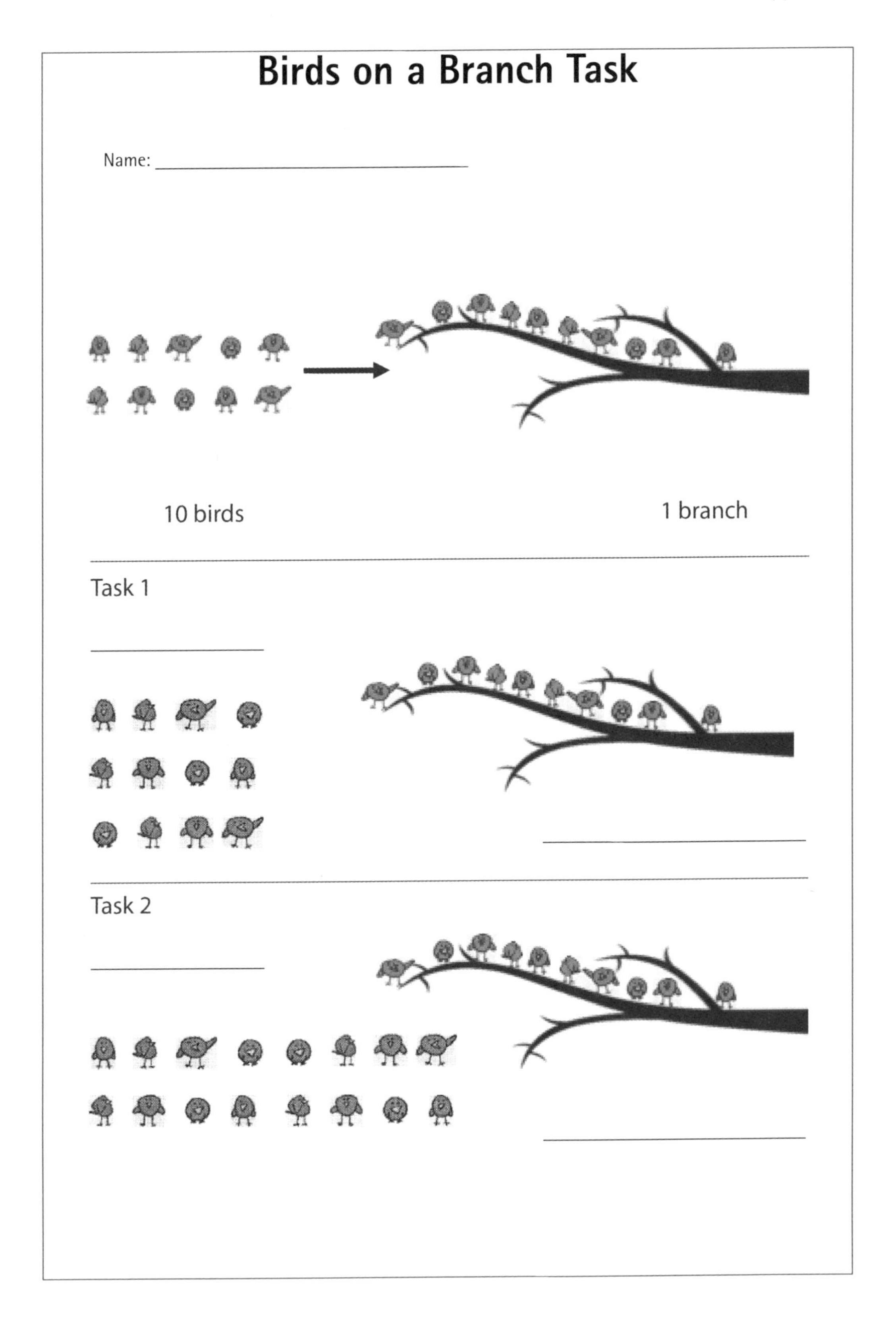
Birds on a Branch Task
Name:
10 birds
1 branch
Task 1
Task 2

Mixed-Up Places

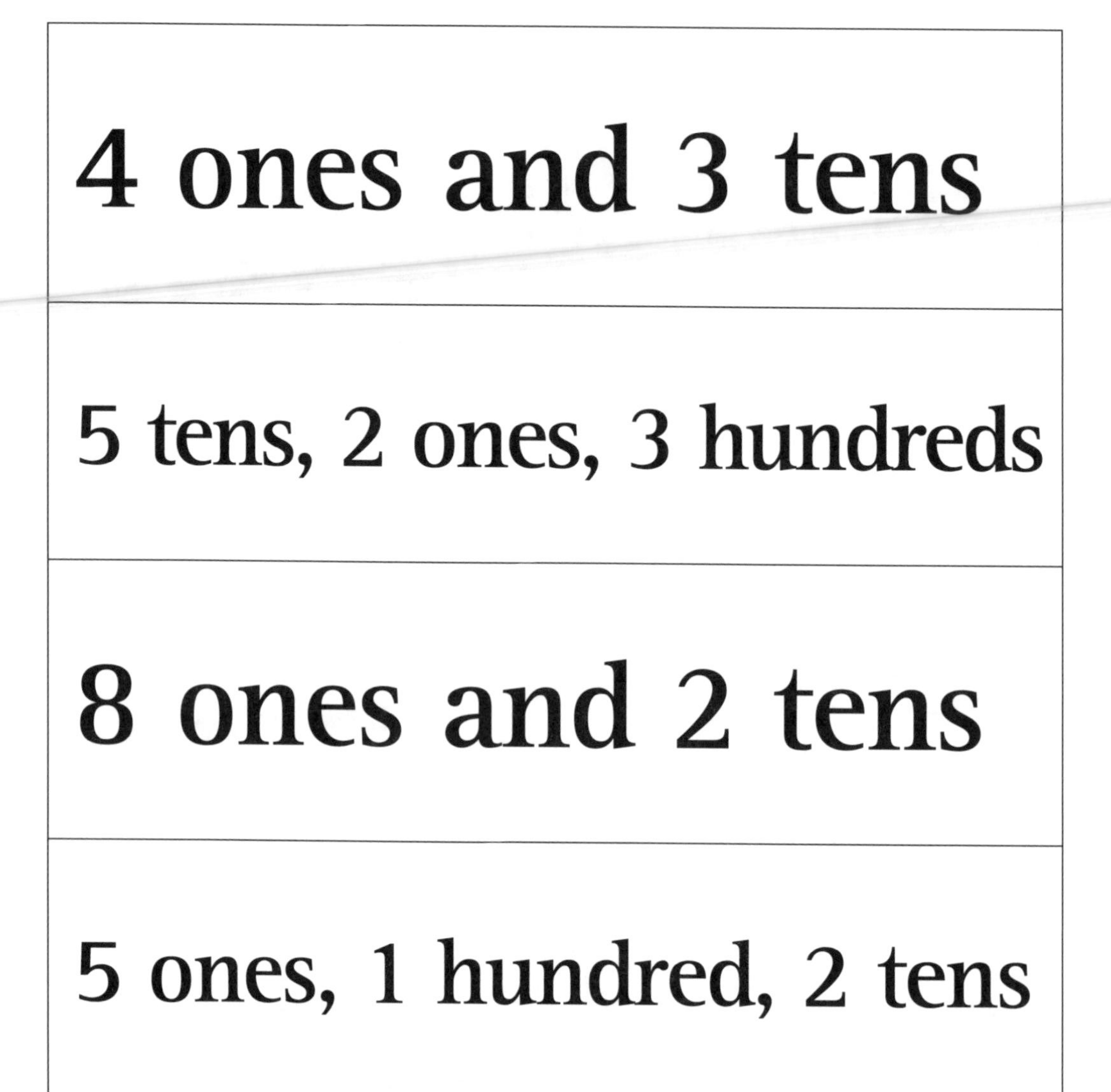

4 ones and 3 tens
5 tens, 2 ones, 3 hundreds
8 ones and 2 tens
5 ones, 1 hundred, 2 tens

Number Stack Cards

1	0	0	0	1
2	0	0	0	2
3	0	0	0	3
4	0	0	0	4
5	0	0	0	5

Number Stack Cards

1	0	0	0	1
2	0	0	0	2
3	0	0	0	3
4	0	0	0	4
5	0	0	0	5

Number Stack Cards

6	0	0	0	6
7	0	0	0	7
8	0	0	0	8
9	0	0	0	9
1	0	0	1	0

Number Stack Cards

2	0	0	2	0
3	0	0	3	0
4	0	0	4	0
5	0	0	5	0
6	0	0	6	0

Number Stack Cards

7	0	0	7	0
8	0	0	8	0
9	0	0	9	0

References

Bashash, Laaya, Lynne Outhred, and Sandra Bochner. "Counting Skills and Number Concepts of Students with Moderate Intellectual Disabilities." *International Journal of Disability, Development and Education* 50 (September 2003): 325–45.

Berteletti, Ilaria and James R. Booth. "Perceiving Fingers in Single-Digit Arithmetic Problems." *Frontiers in Psychology* 6 (March 16, 2015): 226.

Booth, Julie. L., and Robert S. Siegler. "Numerical Magnitude Representations Influence Arithmetic Learning." *Child Development* 79 (July 2008): 1016–31.

Bueti, Domenica, and Vincent Walsh. "The Parietal Cortex and the Representation of Time, Space, Number and other Magnitudes." *Philosophical Transactions of the Royal Society of London Series B: Biological Sciences* 364 (July 2009): 1831–40.

Chan, Becky Mee-yin, and Connie Suk-han Ho. "The Cognitive Profile of Chinese Children with Mathematics Difficulties." *Journal of Experimental Child Psychology* 107 (November 2010): 260–79.

Claessens, Amy, and Mimi Engel. "How Important Is It Where You Start? Early Mathematics and Later School Success." *Teachers College Record* 115 (June 2013): 1–29.

Clements, Douglas H., and Julie Sarama. *Learning and Teaching Early Math: The Learning Trajectories Approach*. 2nd ed. New York: Routledge, 2014.

Comrie, Bernard. "Endangered Numeral Systems." In *Bedrohte Vielfalt: Aspekte des Sprach(en) tods–Aspects of Language Death*, edited by Jan Wohlgemuth and Tyko Dirksmeyer, pp. 203–30. Berlin: Weißensee Verlag, 2005.

Crosby, Martha E., and Catherine Sophian. "From Subitizing to Counting." In *Proceedings of the Human Factors and Ergonomics Society Annual Meeting* 47, pp. 1600–04. Los Angeles: SAGE Publications, 2003.

Dehaene, Stanislas, Véronique Izard, Elizabeth Spelke, and P. Pica. "Log or Linear? Distinct Intuitions of the Number Scale in Western and Amazonian Indigene Cultures." *Science* 320 (March 2008): 1217–1220. doi:10.1126/science.1156540

Dougherty, Barbara J. "Access to Algebra: A Process Approach." In *The Future of the Teaching and Learning of Algebra*, edited by Helen Chick, Kaye Stacey, Jill Vincent, and John Vincent, pp. 207–13. Melbourne, Victoria, Australia: University of Melbourne, 2001.

Dougherty, Barbara. "A Quantitative View of Early Algebra." In *Algebra in the Early Grades,* edited by James J. Kaput, David W. Carraher, and Maria L. Blanton, pp. 389–412. New York: Lawrence Erlbaum Associates, 2008.

Dougherty, Barbara J., Alfinio Flores, Everett Louis, Catherine Sophian, and Rose Zbiek. *Developing Essential Understanding of Number and Numeration for Teaching Mathematics in Pre-K–2*. Reston, Va.: National Council of Teachers of Mathematics, 2010.

Dougherty, Barbara J., and Linda C. Venenciano. "Measure Up for Understanding: Reflect and Discuss." *Teaching Children Mathematics* 13 (May 2007): 452–456.

Dowker, Ann, Sheila Bala, and Delyth Lloyd. "Linguistic Influences on Mathematical Development: How Important Is the Transparency of the Counting System?" *Philosophical Psychology* 21 (August 2008): 523–38.

Ebersbach, Mirjam, Koen Luwel, and Lieven Verschaffel. "The Relationship between Children's Familiarity with Numbers and their Performance in Bounded and Unbounded Number Line Estimations." *Mathematical Thinking and Learning* 17 (June 2015): 136–54.

Elkind, David. "Piaget's Conservation Problems." *Child Development* (March 1967): 15–27.

Engel, Mimi, Amy Claessens, and Maida A. Finch. "Teaching Students What They Already Know? The (Mis)alignment between Mathematics Instructional Content and Student Knowledge in Kindergarten." *Educational Evaluation and Policy Analysis* 35 (June 2013): 157–78.

Feigenson, Lisa, Stanislas Dehaene, and Elizabeth Spelke. "Core Systems of Number." *Trends in Cognitive Sciences* 8 (July 2004): 307–14.

Fischer, Martin. H. "Finger Counting Habits Modulate Spatial-Numerical Associations." *Cortex 44* (April 2008): 386–392.

Fuson, Karen C. "Conceptual Structures for Multiunit Numbers: Implications for Learning and Teaching Multidigit Addition, Subtraction, and Place Value." *Cognition and Instruction* 7 (December 1990): 343–403.

———. "Avoiding Misinterpretations of Piaget and Vygotsky: Mathematical Teaching without Learning, Learning without Teaching, or Helpful Learning-Path Teaching?" *Cognitive Development* 24 (January 2009): 343–61.

Geary, David C. "Cognitive Predictors of Achievement Growth in Mathematics: a 5-year Longitudinal Study." *Developmental Psychology* 47 (November 2011): 1539–52.

Geary, David C., Christine Bow-Thomas, Liu Fan, and Robert S. Siegler. "Even before Formal Instruction Chinese Children Outperform American Children in Mental Addition." *Cognitive Development* 8 (October 1993): 527–29.

Gersten, Russell, Sybilla Beckmann, Benjamin Clarke, Anne Foegen, Laurel Marsh, Jon R. Star, and Bradley Witzel. *Assisting Students Struggling with Mathematics: Response to Intervention (RtI) for Elementary and Middle Schools.* Washington, D.C.: National Center for Education Evaluation and Regional Assistance, Institute of Education Sciences, January 2009.

Gerstmann, Josef. "Syndrome of Finger Agnosia, Disorientation for Right and Left, Agraphia and Acalculia: Local Diagnostic Value." *Archives of Neurology & Psychiatry* 44 (August 1940): 398–408.

Gifford, Sue. "A Good Foundation for Number Learning for Five-Year-Olds? An Evaluation of the English Early Learning Numbers' Goal in the Light of Research." *Research in Mathematics Education* 16 (September 2014) doi: 10.1080/14794802.2014.895677

Gracia-Bafalluy, Maria, and Maria-Pascale Noël. "Does Finger Training Increase Young Children's Numerical Performance?" *Cortex* 44 (April 2008): 368–75.

Grossman, Pamela. *The Making of a Teacher.* New York: Teachers College Press, 1990.

Hamdan, Noora, and Elizabeth A. Gunderson. "The Number Line is a Critical Spacial-Numerical Representation: Evidence from a Fraction Intervention." *Developmental Psychology*, (March 2017): 587–96.

Hanich, Laurie B., Nancy C. Jordan, David Kaplan, and Jeanine Dick. "Performance across Different Areas of Mathematical Cognition in Children with Learning Difficulties." *Journal of Educational Psychology* 93 (September 2001): 615–26.

Jordan, Nancy, C., and Laurie B. Hanich. "Mathematical Thinking in Second-Grade Children with Different Forms of LD." Journal of Learning Disabilities 33, no. 6 (2000): 567–78.

Kamii, Constance. *Young Children Reinvent Arithmetic: Implications of Piaget's Theory.* New York: Teachers College Press, 1985.

Kamii, Constance, Barbara A. Lewis, and Sally Jones Livingston. "Primary Arithmetic: Children Inventing Their Own Procedures." *Arithmetic Teacher* 41 (December 1993): 200–204.

Koponen, Tuire, Kaisa Aunola, Timo Ahonen, and Jari-Erik Nurmi. "Cognitive Predictors of Single-Digit and Procedural Calculation and Their Covariation with Reading Skill." *Journal of Experimental Child Psychology* 97 (July 2007): 220–41. doi:10.1016/j.jecp.2007.03.001

Koponen, Tuire, Paula Salmi, Minna Torppa, Kenneth Eklund, Tuija Aro, Mikko Aro, Anna-Mija Poikkeus, Marja-Kristina Lerkkanen, and Jari-Erik Nurmi. "Counting and Rapid Naming Predict the Fluency of Arithmetic and Reading Skills." *Contemporary Educational Psychology* 44 (January 2016): 83–94. doi:10.1016/j.cedpsych.2016.02.004

Lamon, Susan J. "Ratio and Proportion: Connecting Content and Children's Thinking." *Journal for Research in Mathematics Education* (January 1993a): 41–61.

———. "Ratio and Proportion: Children's Cognitive and Metacognitive Processes." In Rational Numbers: An Integration of Research, edited by Thomas P. Carpenter, Elizabeth Fennema, and Thomas A. Romberg, pp. 131–156. London: Routledge, 1993b.

———. "The Development of Unitizing: Its Role in Children's Partitioning Strategies." *Journal for Research in Mathematics Education* (March 1996): 170–93.

Magnusson, Shirley, Joseph Krajcik, and Hilda Borko. "Nature, Sources, and Development of Pedagogical Content Knowledge for Science Teaching." In *Examining Pedagogical Content Knowledge*, edited by Julie Gess-Newsome and Norman G. Lederman, pp. 95–132. Dordrecht, The Netherlands: Kluwer Academic, 1999.

Moschkovich, Judit. "Revisiting Early Research on Early Language and Number Names." *Eurasia Journal of Mathematics, Science and Technology Education* 13 (June 2017): 4143–56.

National Governors Association Center for Best Practices and the Council of Chief State School Officers Common Core State Standards Initiative (NGA Center and CCSSO). *Common Core State Standards for Mathematics* (*CCSSM*). Washington, D.C.: NGA Center and CCSSO, 2010.

National Research Council Committee. *Mathematics Learning in Early Childhood: Paths toward Excellence and Equity.* Washington, D.C.: The National Academies Press, 2009.

Ng, Sharon, Sui Ngan, and Nirmala Rao. "Chinese Number Words, Culture, and Mathematics Learning." *Review of Educational Research* 80 (June 2010): 180–206.

Passolunghi, M. Chiarra, Barbara Vercelloni, and Hans Schadee. "The Precursors of Mathematics Learning: Working Memory, Phonological Ability and Numerical Competence." *Cognitive Development* 22 (June 2007): 165–84.

Penner-Wilger, Marcie, and Michael L. Anderson. "The Relation between Finger Gnosis and Mathematical Ability: Why Redeployment of Neural Circuits Best Explains the Finding." *Frontiers in Psychology* 4 (December 2013): 877. doi:10.3389/fpsyg.2013.00877

Piaget, Jean. *The Child's Conception of Number.* New York: Routledge & Kegan Paul, 1952 (Original work published 1941).

———. *The Moral Judgment of the Child.* New York: Free Press, 1965.

Piasta, Shayne B., Christina Yeager Pelatti, and Heather Lynnine Miller. "Mathematics and Science Learning Opportunities in Preschool Classrooms." *Early Education and Development* 25 (May 2014): 445–68.

Popham, W. James. "Defining and Enhancing Formative Assessment." Paper presented at the CCSSO State Collaborative on Assessment and Student Standards FAST meeting, Austin, Tex., October 10–13, 2006.

Purpura, David J., Arthur J. Baroody, and Christopher J. Lonigan. "The Transition from Informal to Formal Mathematical Knowledge: Mediation by Numeral Knowledge." *Journal of Educational Psychology* 105 (May 2013): 453–64.

Purpura, David J., and Christopher J. Lonigan. "Informal Numeracy Skills: The Structure and Relations among Numbering, Relations, and Arithmetic Operations in Preschool." *American Educational Research Journal* 50 (February 2013): 178–209. doi:10.3102/0002831212465332

Rips, Lance J., Amber Bloomfield, and Jennifer Asmuth. "From Numerical Concepts to Concepts of Number." *Behavioral and Brain Sciences* 31 (December 2008): 623–42.

Ross, Sharon H. "Parts, Wholes, and Place Value: A Developmental View." *The Arithmetic Teacher* 36 (February 1989): 47–51.

Ross, Sharon R. "Place Value: Problem Solving and Written Assessment." *Teaching Children Mathematics* 8 (March 2002): 419–23.

Sarama, Julie, and Douglas H. Clements. *Early Childhood Mathematics Education Research: Learning Trajectories for Young Children.* New York: Routledge, 2009.

Sasanguie, D., B. De Smedt, E. Defever, and B. Reynvoet. "Association between Basic Numerical Abilities and Mathematics Achievement." *British Journal of Developmental Psychology* 30 (June 2012): 344–57. doi:10.1111/j.2044-835X.2011.02048.x

Saxe, Geoffrey B. "A Developmental Analysis of Notational C." *Child Development* 48 (December 1977): 1512–20.

Schaeffer, Benson, Valeria H. Eggleston, and Judy L. Scott. "Number Development in Young Children." *Cognitive Psychology* 6 (July 1974): 357–79.

Schneider, Michael, Simon Merz, Johannes Stricker, Bert De Smedt, Joke Torbeyns, Lieven Verschaffel, and Koen Luwel. "Associations of Number Line Estimation with Mathematical Competence: A Meta-Analysis." *Child Development* 8 (April 2018): 1–18. doi: 10.1111/cdev.13068

Shulman, Lee S. "Those Who Understand: Knowledge Growth in Teaching." *Educational Researcher* 15, no. 2 (1986): 4–14.

______. "Knowledge and Teaching." *Harvard Educational Review* 57, no. 1 (1987): 1–22.

Siegler, Robert S., and Julie L. Booth. "Development of Numerical Estimation in Young Children." *Child Development* 75 (March 2004): 428–44.

Siegler, Robert S., and Geetha B. Ramani. "Playing Linear Number Board Games—But Not Circular Ones—Improves Low Income Preschoolers' Numerical Understanding." *Journal of Educational Psychology* 101 (August 2009): 545–60.

Sophian, Catherine. "Measuring Spatial Factors in Comparative Judgments about Large Numerosities." In *Foundations of Augmented Cognition* (Third International Conference Proceedings), edited by Dylan D. Schmorrow and Leah M. Reeves, pp. 157–65. Berlin/Heidelberg: Springer-Verlag, 2007. doi: 10.1007/978-3-540-73216-7

Sophian, Catherine. *Children's Numbers*. Madison, Wisc.: Brown & Benchmark Publishers, 1995.

Sophian, Catherine. "Early Developments in Children's Understanding of Number: Inferences about Numerosity and One-to-One Correspondence." *Child Development* 59 (October 1988): 1397–1414.

Sophian, Catherine. "Early Developments in Children's Use of Counting to Solve Quantitative Problems." *Cognition and Instruction* 4 (June 1987): 61–90.

Thompson, Frederick Ian. "Addressing Errors and Misconceptions with Young Children." In *Teaching and Learning Early Number*, edited by Frederick Ian Thompson, pp. 205–13. Maidenhead, U.K.: Open University Press, 2008.

van Klinken, Eduarda, and Emma Juleff. "They Still Can't Count." *Australian Primary Mathematics Classroom* 20 (October 2015): 9–13.

Wiliam, Dylan. "Keeping Learning on Track: Classroom Assessment and the Regulation of Learning." In *Second Handbook of Research on Mathematics Teaching and Learning*, edited by Frank K. Lester, Jr., pp. 1053–98. Charlotte, N.C.: Information Age; Reston, Va.: National Council of Teachers of Mathematics, 2007.

Yinger, Robert J. "The Conversation of Teaching: Patterns of Explanation in Mathematics Lessons." Paper presented at the meeting of the International Study Association on Teacher Thinking, Nottingham, England, May 1998.

Zhang, Xiao, Tuire Koponen, Pekka Räsänen, Kaisa Aunola, Marja-Kristiina Lerkkanen, and Jari-Erik Nurmi. "Linguistic and Spatial Skills Predict Early Arithmetic Development via Counting Sequence Knowledge. *Child Development* 85 (May 2014): 1091–1107. doi:10.1111/cdev.12173